Forensic Enforcement
The Role of the Public Analyst

Dedication

This book is dedicated to the many Public Analysts who over the last 150 years have dedicated their lives to ensuring that food adulteration is not commonplace and that, in the UK in particular, the public can enjoy food which is compliant with legislation, as described on the label, and fit for purpose. Over the last thirty years or so I have met many Public Analysts and all have been keen to develop colleagues and help them progress. The proceeds of this book are going to support this cause and to assist training of future Public Analysts.

The badge of the association (left), has a design which is partly descriptive and partly symbolic, with various motifs in its composition indicating or alluding to some of the more important duties of Public Analysts.

- The ears of corn in the right and the poppy head on the left, representing respectively a typical food and a valuable drug. The examination of foods and drugs may be said to be the *raison d'être* of the Public Analyst.
- In the centre of the design is a mortar and pestle familiar as a device found in every laboratory from time immemorial.
- The Sword of Justice alludes to the protection his labours are intended to afford alike to the all-trusting consumer and the honest vendor.
- The open book may simply be regarded as providing a pictorial allusion to the learned nature of his profession, and its ever open state denotes the daily use he makes of that learning.
- The pendant takes the form of a balance weight in reference to the quantitative aspect of his work. The significance and usefulness of the analyst are dependent on their accuracy.
- His duties embrace the use of gases, liquids and solids. Since his search for the unknown extends into the earth itself; its constituents and its fruits; into the water lying or falling upon it; and into the all pervading atmosphere; these three elements are fittingly depicted by blue representing the atmosphere in which he searches for pollution, by silver, the water of which he ensures purity for drinking purposes, and by red, the earth, the source and origin of most of his raw material.
- The circular shape of the badge alludes to the never-ending character of his work which is directed towards the betterment and protection of his fellow creatures.

Forensic Enforcement
The Role of the Public Analyst

Glenn Taylor

RSCPublishing

ISBN: 978-1-84755-871-8

A catalogue record for this book is available from the British Library

Published by The Royal Society of Chemistry,
Thomas Graham House, Science Park, Milton Road,
Cambridge CB4 0WF, UK

Registered Charity Number 207890

For further information see our website at www.rsc.org

Foreword

The meaning of words and the frequency of their use changes with time. 'Forensic' is in much more common parlance than it was twenty years ago, partly due to the plethora of TV series which have popularised the work of crime scene investigators. The work of Public Analysts, like any other forensic scientist, is rarely glamorous but is an essential part of the legal process. There have been no major miscarriages of justice where a Public Analyst's evidence has been questioned: unlike fingerprints or paediatrics our work has never suffered from much publicity, good or bad.

In the United Kingdom food law is criminal law: a factory, shop or pub can be a crime scene. In an industry worth, some estimate, £150 billion annually, food crime can be a lucrative endeavour; from spirit substitution in pubs and clubs to passing run-of-the-mill food off as organic there is money to be made. Yet, as you will read, there is little money in enforcement analysis. Techniques such as those involving high resolution mass spectrometry, for example, peptide sequencing or dioxin analysis, are beyond the reach of local authorities and, in the current economic climate, are likely to remain so.

The Society of Public Analysts was set up in 1874. Its first tasks were to define adulteration and to promote mutual assistance and co-operation among Public Analysts. It is ironic that today legislation is filled with definitions and despite a wish to cooperate we are discouraged from doing so by a need to compete for ever-diminishing numbers of samples. There were four Public Analysts from Yorkshire at the first meeting. Today that would require a full turnout. But a small number of Public Analysts is not a problem in itself. The problem is one of lack of strategic direction. Fifty years ago Dr Lewis Coles (Public Analyst for Mid and West Glamorgan and President of the Association of Public Analysts 1977–1979) saw the only answer to what were then seen as problems of falling salaries and prestige was 'regionalisation'. He wrote, 'such an idea would mean revolutionary changes . . . . something which could only be decided at central government level'. Laboratories have closed due to local political and financial decisions, not strategic ones. In June 2009, Dawn Primarolo, then Minister of State for Public Health, said "it is absolutely imperative to maintain the highest standard of food control". Ministers move on, the financial climate and governments change, but the pubic have a right to expect an effective, risk-based and proportionate enforcement system which, in my opinion, must include adequately resourced and equipped laboratories staffed with sufficient qualified and experienced staff. At present there is no mechanism for ensuring the continuation of laboratories, let alone maintaining them at the highest standard.

Dr Duncan Campbell
President, Association of Public Analysts

Preface

Who came first, the forensic scientist or the Public Analyst? Probably a pointless argument; there is solid evidence that chemistry was in use well before the 18th century for the detection of both the adulteration of food and water, and murder. In fact, forensic means 'pertaining to or used in a court of law'; from the Latin *forensis* meaning before the forum. Therefore Public Analysts *are* forensic scientists and were granted some of the first laws entitling them to present evidence to court regarding the detection of fraud and adulteration. They use their chemical and general scientific knowledge and present evidence in court in support of enforcement officers including Trading Standards, Environmental Health, Customs and Police officers in their duties to solve and prosecute crime. Certainly Public Analysts were early forensic scientists. The 1860 Adulteration of Food Act was one of the first Acts to require the use of science to prove adulteration.

This book describes some of the history and records the memoirs of a few of the scientists and investigators who together form the UK enforcement team. Environmental Health Officers, Trading Standards officers and Public Analysts work together to protect the public and, over the last 150 years, have led the fights towards keeping healthy and against fraud, whether or not it resulted from a deliberate act or an accident.

Forensic Enforcement: The Role of the Public Analyst
By Glenn Taylor
© Glenn Taylor 2010
Published by the Royal Society of Chemistry, www.rsc.org

From the founding of enforcement, by scientist Carl Friedrich Accum in the early 1800s, to the present day, it has been necessary for enforcers to be diligent, single-minded, resourceful and ingenious. Finding evidence to prove a case or developing tests to detect adulteration or fraud and subsequently, in the true sense of science, sharing these methods with colleagues so they can be questioned, was a practice employed widely to provide public protection and further develop and improve the detection of fraudulent activity.

The focus against fraud has not always been as sharp as it might have been. The 1990s in particular saw the focus shift towards scientist competing against scientist in the quest for best value. This was a term employed at the time to help the government demonstrate that it was spending its scarce funds wisely. The Best Value initiative was instigated to introduce competition between suppliers of services to government departments. As a result, some scientific service providers (who were not trained business people and consequently often lacked business skills) became quasi-businesses searching for a competitive advantage; and as a consequence were reticent in sharing information freely. Gone are the days of sharing methods with colleagues so that they can be questioned, improved and developed. Research and development became an in-house secret. Scientists were not happy to allow a neighbouring laboratory to develop in an area complementary to their own as the neighbours might gain commercial advantage. Survival was all. Many business gurus might argue for Darwin's evolutionary principles or 'survival of the fittest' in the business sense to be applied to this group of scientists to inspire innovation, arguing that it would bring out the best in enforcement. This new regime resulted in enforcers spending time detailing precise contract specifications for their work and preparing detailed tenders for scientists to compete against each other. Typically, a three-yearly ritual of tenders began and scientists learned to find competitive advantage, but lost the benefit of cooperation, to the detriment of national enforcement. Since the 1990s, probably as a consequence of this forced competition, more enforcement laboratories have closed than ever before. Best Value?

Accum is widely accredited as the father of the group of scientists called Public Analysts, a highly skilled and highly qualified group who exist to protect the public by using their scientific expertise.

Much of the work of Public Analysts is related to food. However, this book is not solely dedicated to food-fraud, but demonstrates the range of forensic science skills and investigation employed by this dedicated team and their colleagues in search of the true answer. Traditionally, these scientists have been at the forefront of science to develop tests which enable them to provide independent expert opinion in court, which helps the judiciary to interpret cases and occasionally to form definitive judgements leading to revised regulations. Public Analysts have not always had the support they crave and have definitely ruffled feathers along the way. Always operating in support of enforcement colleagues and thus not often in the limelight, they have rarely been considered first when funding decisions have been made. However, without their opinions, interpretations and skills, evidence would not be complete and prosecutions would not be possible. Without the efforts of all enforcement officers the food-fraud considered rife in the UK in the 1860s might still be prevalent. Does this make them unsung heroes? Perhaps, but they are definitely dedicated to the cause. The advances made against food-fraud have led some, particularly those holding the purse-strings, to consider the scientific fight against fraud to be over and the risks too low to warrant further significant expenditure, leading some to wonder if this group of scientists is nearing extinction. This book is not intended as an argument for the cause. Neither is it a cry for more funds. It simply details some of the events leading to the formation of this group of enforcers and describes some of the work they have undertaken over their 150-year history. It hopefully shows how sophisticated the fraudsters have become and that the fight continues. It needs to, given the global possibilities for fraud and adulteration.

Glenn Taylor

Contents

Forensic Enforcement: The Role of the Public Analyst
By Glenn Taylor
© Glenn Taylor 2010
Published by the Royal Society of Chemistry, www.rsc.org

CHAPTER ONE

Adulteration and the Challenge for Enforcement Scientists

In the UK we have beaten food adulteration. There is an army of enforcement officers: Trading Standards, Environmental Health and Public Analysts who, amongst other things, fight the fight against food adulteration. In 2007, in an attempt to improve efficiency amongst these local authority staff, the government commissioned a review of the 60 policy areas worked on by enforcement officers including food hygiene and adulteration (composition), led by Peter Rogers, chief executive of Westminster City Council. It concluded that adulteration of food (measured by checking its composition) was not in the top five 'priority areas' in terms of risk and effective use of the resource provided by enforcement officers.[1] Consequently, some local authorities now argue that food composition is no longer a priority, leading to it receiving scant attention. Food adulteration has been beaten, or at least you would be forgiven for thinking so.

The problem is that adulteration is difficult to define and means different things to different people. Adulteration was defined by Dr Henry Letheby, 1816–76, as *'the act of debasing a pure or genuine commodity for pecuniary profit, by adding to it an inferior or spurious article, or by taking from it one or more of its constituents'.* To some, (Caroline Walker, the distinguished nutritionalist, writer

Forensic Enforcement: The Role of the Public Analyst
By Glenn Taylor
© Glenn Taylor 2010
Published by the Royal Society of Chemistry, www.rsc.org

and campaigner and others), this includes additives such as emulsifiers, synthetic flavourings and colours, which are added to make a food more desirable. Dr Henry Letheby's definition would seem, at least at first glance, to support her opinion. However, additives such as antioxidants (added to foods that contain fats to stop them going rancid), colours, emulsifiers, stabilisers, gelling agents and thickeners, flavourings, preservatives and sweeteners are permitted in foods and controlled by food law; in the eyes of the law, they are not adulterants. If they were classed as adulterants then synthetic foods, such as an instant strawberry dessert that has never been near a strawberry (or any other fruit) or in fact anything containing additives, *i.e.* ready meals, sweets, snacks and numerous other products would be considered adulterated, leaving the public with much less choice and eating only the most 'natural' ingredients (those grown or raised).

The Public Analyst's role is to help the courts decide what is adulterated and what is not; and, yes, the meaning of those words does change as research provides more evidence. For example, Azo-dyes (in some food colours) once permitted are now considered harmful and have more recently been banned. Also Sudan dyes; colourants used in oil solvents and polishes, which were the subject of an emergency order in the EU in July 2003, were found in chilli powder imported from India in 2005 and subsequently added to many foods including pizzas, sauces and ready meals. Sudan dyes are now considered carcinogenic (an agent that promotes cancer), at least if present in sufficient quantities. Another example is the 'Southampton Six': six colours linked by research at Southampton University with increasing ADHD (Attention Deficit Hyperactivity Disorder) behavioural problems in children. The colourings (Sunset yellow (E110), Quinoline yellow (E104), Carmoisine (E122), Allura red (E129), Tartrazine (E102) and Ponceau 4R (E124)) are now being avoided voluntarily by industry in products aimed at children and there are calls to ban them by law.

On the subject of Public Analysts and enforcement officers, Professor Tim Lang at City University says: *'It is an enormous job and there are several constraints on their work. Unsurprisingly, one problem is money. Their budgets are squeezed all the time. That's been a long-running concern for me. The real issue isn't about controls, though. It's about the need for more independent food scientists. In the mid-19th century, modern food chemistry began by acting on the people's behalf. What we've got now is chemistry that, by and large, works for the food industry. People in mainstream public health say the problem is not adulterated food, it is heart disease, cancer and diabetes, and they are absolutely right. But the point people like me have been making for years is that the modern legalised adulteration and legalised contamination of food is what enables foods full of hidden fats and sugars to be sold looking like real food. That is the flipside of the coin'.*[2] Professor Lang supports the argument presented by the Food Standards Agency that there is a role for enforcement officers to provide scientific evidence for consumers so that they can make informed choices about the food and the additives that they eat. Given the 'obesity epidemic' and the associated risks to health from food, particularly processed foods, it is hard not to conclude that food standards should be a high priority for enforcement. Obesity alone costs the NHS about £5 billion per year rising to £10 billion by 2050. Obesity is responsible for 9000 premature deaths each year in England and reduces life expectancy on average by nine years.[3] Perhaps Peter Rogers *et al.* should think again? Food composition should be a high priority for enforcement. Maybe then some of the local authorities who argue that they should spend their efforts elsewhere would renew their vigour in this area and seek to provide the solid scientific evidence for their residents and tax payers.

So, what is adulteration? When I was at university, one of the inorganic chemistry lecturers was often heard stating: *'everything is toxic; it's just a question of the amount needed to kill'.* So, should we adopt a 'precautionary approach' and ban all additives until proven absolutely safe? Maybe Caroline Walker, Tim Lang and others have a point, but, for now, we have to go with the current legislation and keep up the research.

Do we have food adulteration and fraud under control in the EU? Recent food scares such as the 'Surrey Curry Scandal'[4] where chicken tikka masala investigated by Surrey Trading Standards Officers was found to contain up to four times the limit for permitted colours; the despicable addition of melamine to Chinese baby milk[5] to produce a product sufficiently high in apparent protein to pass off as milk, but which had no nutritional value and led to renal failure and malnutrition in infants; the Sudan[6] red dye in spices; and counterfeit vodka,[7] would certainly suggest not – the fight continues.

Is adulteration new? In 250 BC King Hiero commissioned a gold crown. He gave the goldsmith the gold and when he received the new crown he was concerned that it had been adulterated. He thought that the goldsmith had kept some of the gold for himself and replaced some of the gold with silver. King Hiero asked Archimedes to investigate the matter but without damaging the crown. At the time no techniques existed for an analysis of this kind without causing damage. A perplexed Archimedes had to develop his own test. Whilst taking a bath, Archimedes had the 'Eureka moment' when he discovered that his body displaced a consistent volume of water in the bath. He realised that he could use this to measure the density of the gold crown and discover if adulteration had indeed taken place. He was, it is said, so excited by the discovery that he ran down the street naked shouting '*Eureka*' (I have found it!). He checked the density of the crown and compared it to genuine gold and silver. It is alleged that this demonstrated that adulteration had indeed taken place. This story was recorded by Vitruvius some 200 years later and has been doubted by some scholars who argued that Archimedes would have had to make precise measurements of the crown's volume, which had an irregular shape; something that would have been incredibly difficult at the time. However, the total volume of water displaced is equivalent to the volume of the crown and therefore accurate measurements of density and volume could have been made.[8]

There are several references to adulteration of wine and bread which trace the practice back as far as the Greeks and Romans. Pliny (AD70) recorded 'the wheat of Cyprus is swarthy and produces dark bread, for which reason it is generally mixed with white wheat of Alexandria'. He noted, with disapproval, that some bakers kneaded their bread with seawater, so that they could save on the cost of salt and detailed how white earth 'Leucogee' (alum, associated more latterly with dementia) was added to bread to bulk its weight.

In 1202, King John declared an 'Assize of Bread', in an attempt to control the profitability of bakers and fix the price of bread to the price of wheat. By linking the price of bread to the price of wheat King John hoped to stave off famine and the associated uprisings that normally followed. Sadly an attempt at controlling profitability did no such thing. It merely encouraged profiteering by other means, namely adulteration. The bread was sold by weight and thus the fraudulent bakers simply increased the weight of the produce by using sawdust or metal amongst other things. Consequently the Assize had to be regularly updated to include regulations concerning the adulteration; as methods of adulteration changed, so did the Assize. In 1582 the Assize, which had effectively become the English Sale of Food Act, included details of punishments for first, second, third and fourth offences. This resulted in a fine, loss of stock, pillory or prison and, finally, banishment from town.

In 1311, Alan de Lyndseye, a baker, was brought before the Alderman and Mayor and 'sentenced to the pillory for making bread that was of bad dough within and good dough on the outside'. He didn't seem to learn. A short while later he was back before the Alderman and Mayor and again sentenced to the pillory for 'selling bread that was made of false, putrid and rotten materials through which those who brought the bread were deceived and might be killed'.[9]

Beer was also a key part of the diet and therefore a target for profitable adulteration. The quality of drinking water could not be relied upon and therefore beer consumption was encouraged. It also provided additional nutritional value when diet might not have been completely 'balanced'. At the end of the 17th century a child was allowed to drink two pints per day. Benjamin Franklin, who lived in London during the period 1757–1774, recorded the daily beer consumption in a London printing house which he visited. The employees each had a pint before breakfast, a pint between breakfast and dinner, a pint at dinner, a pint at six o'clock and a pint when they finished work.

In the 1700s, people recognised food was adulterated. Tobias Smollett, a physician, author and satirist wrote in *The Expedition of Humphry Clinker* in 1771:

> 'The bread I eat in London is a deleterious paste, mixed up with chalk, alum and bone ashes, insipid to the taste and destructive to the constitution. The good people are not ignorant of this adulteration; but they prefer it to wholesome bread, because it is whiter than the meal of corn [wheat]. Thus they sacrifice their taste and their health to a most absurd gratification of a misjudged eye; and the miller or the baker is obliged to poison them and their families, in order to live by his profession.'

The Victorians were sticklers for standards and rules, although clearly not all Victorians followed them. During the 18th and 19th centuries food adulteration was rife (Figure 1.1). Strychnine was used to make beer taste bitter, alum (now possibly linked to dementia) and chalk were used to increase the brightness of flour, and weight was increased using sawdust or parsnip powder. Red lead was added to sweets to make them brighter and more colourful.

Two London based scientists, Accum and Hassall, led the charge to stop this fraud.

Figure 1.1   Little girl: '*If you please, Sir, Mother says, will you let her have a quarter of a pound of your best tea to kill the rats with, and an ounce of chocolate as would get rid of the black beetles*'. Punch 14 August 1855.

In 1820, Frederick Accum raised the alarm on food adulteration. His analytical work led him to publish a book: '*A Treatise on Adulterations of Food and Culinary Poisons*', the first edition of which sold out within a month, demonstrating public awareness and concern. This book is still available today.

Between January 1851 and December 1854, Hassall analysed around 2500 samples in his London laboratory and proved that adulteration was the rule, not the exception. He recorded the names and addresses of the vendors and published these details along with the results of his analysis. Hassall used microscopy and rudimentary chemical tests to identify food-fraud. Microscopy was introduced as a means of identifying coffee adulteration. Coffee had become very popular by 1850 and was relatively expensive thus making it a target for fraudsters. Coffee was 'cut' with chicory in the same way that a drug dealer might 'cut' or dilute illegal drugs today. Whilst chicory is not harmful to health the practice of 'cutting' made vast sums of money for the fraudsters.

The Adulteration of Food and Drink Act 1860 was introduced in an attempt to reduce the levels of adulteration. It made provision for the appointment of Public Analysts; qualified scientists dedicated to the protection of the public. Not all local authorities appointed Public Analysts. Some decided that the money should be spent elsewhere. The 1860 Act was revised in 1872 to incorporate Hassall's proposals including the naming and shaming of those found guilty of adulteration (now it was legal to provide this information) and for a second offence, punishment by six months' hard labour. Local authorities were required to appoint Public Analysts and the full-scale fight against food-fraud began. At this time, Hassall was seen as one of the leading lights in food adulteration and its detection.

In 1874, the Society of Public Analysts was founded. Hassall was instrumental in its formation and was involved in examining the working of the 1872 Act. Hassall again gave evidence and the report of his committee provided the basis for the following Acts: Sale of Food and Drugs Act of 1875 (amended 1879); the Margarine Act of 1887; and the Food Adulteration Act of 1899. These Acts produced considerable improvements in food standards. Hassall's investigations eventually resulted in dramatic improvements in the control of adulteration, the appointment of Public Analysts in all the counties and boroughs of Britain and the formation of the Society of Public Analysts to represent their interests and maintain their professional status. Thus the first model of enforcement was established in 1874 with 77 Public Analysts, each with his own laboratory, working together through a society to provide uniform methods of analysis and common interpretation of analytical results.

The laws for the prevention of adulteration of food, drink and drugs in 1872 did not strictly define adulteration but declared that the admixture of anything whatever with an article of food, drink, or drug, for the purposes of fraudulently increasing its weight or bulk, is an adulteration within the previous provision of the Act. The adulteration of intoxicating liquors was covered in the Licensing Act 1872, which provided a schedule of deleterious ingredients which were considered to be adulterations. Adulterants of interest to enforcement are manifestly of a fraudulent nature and whilst some may arise accidentally or naturally in foods and other commodities the tracing of the source is of key interest in pursuing a charge of adulteration.

In 1898 the then Institute of Chemistry qualified Public Analysts who had to pass an examination in the chemistry, microbiology and microscopy of food, water and agricultural fertilisers and feeding stuffs. This examination eventually became the Branch E of the Fellowship Examination, and then the Mastership of Chemical Analysis of the Royal Society of Chemistry. It is incorporated in the Food Safety (Sampling and Qualifications) Regulations 1990 as the qualification certifying the competence of Public Analysts.

By the mid 1950s, there were 40 laboratories employing over 100 officially appointed Public Analysts. The next two decades saw the introduction of more complex techniques in laboratories, including atomic absorption spectrophotometry for metals analysis and gas–liquid chromatography. Public Analysts became able to identify the additives found in the new 'synthetic foods' encountered in the 1960s and 70s; lemonade made from flavourings and not lemons, being one example of a synthetic food.

Initially, the range of work undertaken by the Public Analyst included food, drugs and water samples. This increased by the 1970s to include: agricultural analysis (animal feeding stuffs, herbicides and pesticides, *etc.*), toxicology, environmental science, health and safety testing and consumer safety (toy safety testing, for example).

The 1980s was a significant decade for Public Analysts and the food industry. This decade saw the introduction of:

- growing numbers of ready meals containing more additives, making food analysis increasingly complex
- changes in lifestyle including microwave cooking leading to significant changes in diets, promoting a 'fast-food culture'
- readily available scientific techniques including high pressure liquid chromatography (HPLC), Fourier transform infrared spectrometry (FTIR), immunoassays and enzymatic analysis
- externally assessed laboratory quality systems and accreditation (by NAMAS, now UKAS) designed to ensure the quality and consistency of analysis and the interpretation of results
- externally assessed proficiency trials, designed to test the ability of the laboratory and its scientists to analyse unknown samples accurately and precisely; and
- Local Government Act 1988 which introduced compulsory competitive tendering to local authorities.

Most of the initiatives introduced in the 1980s and implemented in the 1990s had a significant upward impact on the costs of Public Analysts, who embraced the new rules despite this. The costs of the improved technology and quality systems (the latter estimated to add 10–20% to laboratory costs) were very significant. Whilst compulsory competitive tendering did not apply directly to Trading Standards or Public Analysts, the 1990s saw the introduction of tendering for services, pitting one analyst against another and driving laboratory fees (but not costs) down. This resulted in some Public Analysts losing or having to share their long-held appointments with others, leading to reductions in income. This started the search for competitive advantage and reduced the likelihood of working together for the benefit of enforcement.

The 1999 Food Standards Act established a new model for food enforcement in the UK; one which included Environmental Health (food hygiene), Trading Standards (food composition), Public Analysts, and LACORS (Local Authority Coordinators of Regulatory Services), all representing local authorities. The newly formed (2000) Food Standards Agency (FSA) represented central government as the regulators. The Act gave the FSA powers to (*inter alia*):

- set standards of performance in relation to enforcement of food law; and
- monitor the performance of local food enforcement authorities.

An analysis of expenditure between 2002–2008 as recorded by CIPFA (Chartered Institute of Public Finance and Accountancy) shows a drop of about 30% (no inflation allowance) in expenditure by local authorities in England with their in-house Public Analyst laboratory. However, many authorities helped their laboratories survive with additional work/subsidies behind the scenes. English authorities without their own laboratories but who use the services of outside PA services have dropped their expenditure by around only 3% (again no inflation allowance) over the same period. However, an analysis of tenders over a ten-year period shows a different story, with a significantly reduced expenditure on food law enforcement analysis of around 50%. It can therefore be seen that total expenditure on food standards by UK local authority enforcement has dropped significantly, especially when inflation is considered. As Public Analysts tender for work against each other, the pressure is on them to reduce their fees in order to win the work. This leads to less money being available to be spent on training and new technology within the laboratories. Consequently it is harder for the FSA to ensure that adequate provision is available, especially for new food concerns such as genetically modified (GM) foods and other matters as they arise. Melamine in foods is one of the latest examples (the melamine issue is discussed in Chapter 6).

In the last ten years there have been nine laboratory closures and with reducing resources and low investment in some local authority run laboratories it seems that this service (which is a vital part of the English food safety system) is approaching a crisis point. Currently there are 18 laboratories employing 41 Public Analysts, 27 (66%) of whom are over the age of 50. The reduction in the service causes a reticence for new scientists to undergo the rigorous training and qualification required to become a Public Analyst and it is foreseeable that there will not be sufficient new Public Analysts to maintain the current system.

Public interest in food and diet has never been higher. New issues such as GM foods, novel foods (foods which have recently become in vogue not having been consumed in significant amounts previously), health related products and organic food concern the public. Whilst old-style food-fraud and adulteration is no longer 'rife' as a result of the efforts of the enforcement profession, around 40% of samples submitted to Public Analyst laboratories still fail to meet legislative standards. Approximately 1 in 3 fail food labelling requirements and 1 in 10 fail compositional regulations. The model for food law enforcement, which was established nearly 150 years ago, has served the public well, reducing food-fraud, although food adulteration is still encountered.

This book is not an argument for banning additives, but focuses on those compounds which have been added, perhaps by mistake and occasionally fraudulently, and have been found in samples submitted to the laboratory for analysis. It focuses on a by-product of adulteration – the Public Analysts – a group of scientists seldom heard who have dedicated their careers to the proof of fraudulent activity and the education of consumers based on sound scientific evidence. It is their duty to protect consumers against adulteration produced by whatever means and to assist the courts in deciding whether or not an offence has been committed. They advise on whether or not it was harmful or whether it was prejudicial to the consumer. Without adulteration there would be no need for enforcement. Adulteration impacts on all consumables and consequently this book will not solely relate to food, but will include other examples, demonstrating the wide range and complexity of work undertaken in Public Analyst laboratories and their equivalents throughout Europe.

REFERENCES

1. *National Enforcement Priorities for Local Authority Regulatory Services*, March 2007, ISBN: 0-7115-0479-2.
2. S. Kinnes, *Guardian* newspaper, Saturday May 15, 2004.
3. http://www.dh.gov.uk/en/Publichealth/Healthimprovement/ Obesity/DH_078098
4. S. Kinnes, *Guardian* newspaper, Saturday May 15, 2004.
5. http://www.fsascience.net/2008/09/24/chinese_milk
6. http://cot.food.gov.uk/pdfs/TOX-2003-41.PDF
7. http://www.food.gov.uk/news/newsarchive/2008/sep/vod
8. W. Blyth and H. Cox, *Foods; Their Composition and Analysis*, Charles Griffin and Co. Ltd, 1927.
9. H.T. Riley, *Memorials of London*, pp 120, 121.

CHAPTER TWO

Nature or Nurture; are Scientists Born or Manufactured?

'Difficult to deal with', 'tenacious' and 'bloody-minded' are three of the 'compliments' I have heard used to explain the characteristics of scientists in general and in particular of Public Analysts. At the time they probably weren't intended as a compliment! I expect words such as fastidious, demanding and exacting would be considered by many scientists as compliments and something to aspire to. Clearly not everyone in the profession is like that. I have worked with some very persuasive individuals, who will not be named here, but I have often wondered why Public Analysts might be seen as robust or difficult to deal with and single minded. In researching this book I began to understand more about their training and character. This led me to wonder if they are born that way and attracted to the profession or if perhaps it is a necessary quality acquired through training. Most probably it is a mixture of the two. This chapter doesn't seek to answer this question, but in some cases it does demonstrate what kind of person is attracted to this area of science and I suspect it will throw some light on why.

Forensic Enforcement: The Role of the Public Analyst
By Glenn Taylor
© Glenn Taylor 2010
Published by the Royal Society of Chemistry, www.rsc.org

CELEBRITY STATUS

The 19th century was a period of major change; knowledge in science and medicine improved dramatically, ailments which were fatal became treatable, adulteration previously undetected became detectable and books written by physicians who were often also chemists sold out. This was especially true for those relating to the detection of food-fraud, which became a key issue for the public. Consequently, many scientists attained celebrity status, their lectures being widely attended and the innovations widely reported. Some may say that it is sad that this status is not conferred on scientists today. I for one!

> In *'The Fight Against Food Adulteration'* Lawson Cockroft writes: *'Accum published his "System of Theoretical and Practical Chemistry" to support his teaching activities in 1803. At the time chemistry was then a very fashionable subject and the people who bought the book or went to the lectures were leisured men and women with time on their hands. Accum was a great believer in setting up demonstrations and in getting his pupils to do their own experiments. His series of "Chemical Demonstrations or Private Lectures" proved highly popular and, for a long time, his laboratory was the only institution where instruction in practical chemistry was offered. His advertisements for the courses were very persuasive and claimed that skill could only be acquired by actual practice. To achieve this, he offered private, as well as public lectures and took resident pupils'.*[1]

Dr Arthur Hill Hassall was one of the first Public Analysts and in 1872 advised the government on legislation to protect the public against food-fraud. He was also a physician who established the Royal National Hospital for Diseases of the Chest in Ventnor to treat tuberculosis, which at the time required convalescence because medicines were unavailable. He treated, amongst others, Charles Dickens, Karl Marx and Alfred Lord Tennyson.[2]

Perhaps one of the more popular scientists of the time, John Pepper, established science (physics) as part of the theatre, basing his live experiments on the work of William Brande and Michael Faraday. Pepper's skill was the ability to take the experiments of others and demonstrate them in a spectacular manner. As part of his repertoire he took a Faraday experiment and made it spectacular and visible by using a gigantic version of Faraday's induction coil: he produced a spark nearly one metre long to the delight, and probably fright, of 19th century audiences.[3]

THE CHARACTERS

In *'Final Years and Scandal in the Library'*, Cockroft writes: *'A few months after the appearance of the second edition of "A Treatise on Adulterations of Food and Culinary Poisons", a complaint was made to the managers of the Royal Institution, alleging that Accum had been mutilating books in the Institution's library. It was claimed that, though this practice had been going on over a period of years, only recently had evidence appeared linking Accum to the missing pages. At first, the managers were inclined to ignore the charge, especially as Accum had at one time served as their librarian. Further complaints led to Accum being visited again in December 1820 by the librarian and two officers, armed with a search warrant. The librarian identified some thirty pages as being the property of the Royal Institution and Accum was arrested and charged with robbery. He was discharged when the magistrate took the view that though the books might have been valuable, the separate leaves found in Accum's premises were only waste paper. The Royal Institution managers then brought an indictment on the grounds of mutilation of books in their library and a further trial was set for April 1821. Public opinion turned against Accum and public appeals on his behalf by his friends and supporters were in vain. Accum became severely depressed, failed to appear at his trial and forfeited his bail. He remained in London only a few months longer to wind up his affairs and returned, at the age of 52, to his native Germany. He remained in Berlin until his death on 28th June 1838, so deeply affected by the scandal of 1820 that articles which he contributed to the Berlin Royal Academy of Sciences were published either anonymously or under the pseudonym Mucca. Even his London publishers producing new editions of his works omitted his name from their title pages'.*[4]

One wonders if this action was born out of a desire simply to obtain information as quickly as possible; clearly no one would want to condone such action but as a prolific writer perhaps Accum simply focussed on the need to obtain the necessary information to enable him to publish another text. Science is, after all, based on publication and peer review.

During the 1880s Sir George Turner, the Portsmouth Public Analyst, raised concerns that the levels of lead found in local drinking water were too high and consequently the water was harmful to health. His concern led to a court hearing which could not be resolved, possibly due to the fact that there was insufficient supporting material that lead in water, at the levels encountered, could be considered injurious to health. During the court proceedings the managing director of the local water company, who used lead-based pipes to deliver the water, offered Turner the opportunity to settle the matter outside of the court following the 'Queensbury Rules' (a boxing match). It is not detailed as to whether this took place, but nowadays Public Analysts would presumably prefer not to resolve cases in this manner! Despite this I have encountered one or two who have tried to be very persuasive, in a manner not too dissimilar.[5]

Tenacious, fastidious, Public Analysts around at the turn of the 20th century spent years collating data to support their cases and provide information for others to asses their samples (Figure 2.2). On the watering down of milk; Birmingham Public Analysts Dr Alfred Hill and J.F. Liverseedge undertook a monthly analysis and painstakingly compared the results obtained during the thirty-two years 1899–1930, analysing 51,703 samples in total.[6] They were not alone in this level of commitment to the task; Public Analysts at that time certainly demonstrated tenacity and an eye for painstaking detail and the following demonstrates humour and a few in-jokes regarding milk. The jokes are a little obscure and relate to various indices which were used to determine whether or not milk has been watered. This was presented at the Milk Products Sub-committee of the Analytical Methods Committee of the Society of Public Analysts' luncheon on the 13th February 1936 in commemoration of the conclusion of the work of the Sub-Committee and its 85th meeting (Figure 2.3).

Figure 2.2   Leo Taylor, one of the original Public Analysts in his lab in Wal-
thamstow, London, *c.*1900. Possibly growing bacteria using the
Petri dishes on the bench.

Figure 2.3   Public Analysts at a function similar to, if not the one, detailed in
the text.

The Toast of 'The Cow' proposed by A.L. Bacharach, MA, F.I.C.

I wish to thank you, Sir, our honoured host
For asking me to introduce this toast;
With such a privilege I'll not make free.
I must admit the cow appeals to me
Although some others here know better how
It feels to be appealing to the cow.
She has her limitations, but I'll try
To put the microscope to my blind eye.
Let me forget, if only for this hour, that milk is all too ready to turn sour;
That milk, so every Magistrate now knows, is
Sometimes a source of B. tuberculosis;
That milk, defiant of our great profession,
May show abnormal freezing point depression.
Indeed, although the creature does her best,
Her udder's fluid, it must be confessed,
When requisitioned by the trading human
Is not as perfect as is thought by Newman.
However, Sir, the cow is not to blame
For breaking man-made rules in mankind's game;
From cow's ears no one makes a purse of silk
And who asks Bristol cows for Bristol milk?

Come then, we'll turn our thoughts to subjects merrier;
I'm here to praise the cow and not to bury her.
Without the cow, what would our members read?
'The Analyst' were meagre fare indeed
Lacking cryoscopy's collaboration,
The North of England's cold collation.
Without the cow lactosophers would lose
The balanced periods of Hinks and Hughes
Enshrined in words no chemist may ignore,
The M.P.S Reports from one to four.
Nor do I need remind you that, at least
To some extend, we live upon the beast.
To some extend at least our jobs and fees
Are – I feel sure each one of you agrees –

Determined, as a boat's course by its rudder,
By all the products of the bovine udder.
To some extend the hardly-gotten wealth
Of you custodians of Public Health,
Eldon and Parkes and Hinks – and even More
Sitting at Chemistry's official core –
Is gained by catching out the more unwary
Adulterators lurking in the dairy.
So, Bolton, putting milk into a carton
Extracts the profit Debenham's set his heart on.
Even Hughes, Macara and my worthless self
Would hate to put milk products on the shelf
For us the cow, it may be fairly said,
Provides <u>some</u> of the butter on our bread.

And as to Pelle and Anderson, whose words
Control the destinies of countless herds;
Whose daily labours with the children's drink
Frees it from stuff that should go down the sink
Who guard our cream, our butter and our cheese
What coconut is milkier than these?
I'll say, though it's a truism to utter –
The cow provides <u>them</u> with their <u>bread</u> and butter.
And so, Sir, you with customary skill
Have chosen one of them to fill the bill
Of being coupled with this heartfelt toast
To her who helps us all – but them the most.

Gentlemen, charge your glasses to the brim;
About this toast there must be nothing skim.
Shutting our ears to Hanley's tittle-tattle
Let us give praise to all our normal cattle.
May the sweet cow upon the meadows green
Long chew the cud and swallow carotene.
In spring when hearts are amorously full
May she aim truly and bring down a bull.
May she have strength to bear each glad event
And always maintain her fat-content.
May large shadow ne'er grow less – or lesser!
I give you, gentlemen, the cow – God Bless Her.

Perhaps a suitable description of the character of the early Public Analysts is best portrayed by Alan Turner OBE (who became chief examiner of Public Analysts and author of the 1998 Turner report: '*To review Public Analyst arrangements in England and Wales and to make recommendations on how best to provide the scientific and technical support needed by food authorities in respect of their food law enforcement responsibilities, taking account of the concerns of other interested parties, arrangements in other parts of the UK and EU considerations*').

'*Public Analysts of yore were quite strong characters, bordering in some instances on the eccentric, not least Bill Taylor (William Wilders Taylor) who became appointed Public Analyst in Nottingham in 1937. I worked for Bill in the 1940s, he could have given any Government or company lessons in cost saving. The conditions in the lab were beyond Dickensian, not helped by the awful winters and severe fuel shortages at that time. I recall having to thaw out the benzene (benzene freezes at around 5°C) before we could use it. Benzene is a known carcinogen and no longer used in laboratories today. The water seal in the gas meter in the cellar froze and I had to resort to heating a reasonably large iron plate with a blow lamp as an alternative Bunsen burner, tripod and gauze in order to keep the work going. As the main fume cupboard within the lab was given over to Kjeldhal apparatus used for assessing protein levels in foods we used a fireplace and chimney as a fume cupboard, but only when wind conditions outside were suitable. Not much health, and precious little safety, in those days. When Bill observed Alan using BDH (a chemical supplier) supplied acid washed sand to make a sandbath using old tin lids, Bill guided him down to a small cave hewn out of the rock below the normal cellar beneath the laboratory and proceeded to scrape sand off the walls into a tin lid observing that Nottingham was usefully built on sandstone good enough for sandbaths. Even cigarette packets could be reused as 'Notelets', it was common for instructions to be left for me written on the inside of emptied Player's cigarette packets.*

If Bill had been Chancellor these past few years, the Treasury would never have been caught out. He had his generous side though; he would lend me his season ticket for Trent Bridge from time to time. I suppose Bill had what would be called these days a keen eye for the bottom line'.

Throughout the 'troubles' the Public Analyst's laboratory in Belfast city centre was not immune from the general unrest. For Northern Ireland the year 1972 remains tragically notorious as that in which the greatest number (almost 500) lost their lives in incidents related to the 'troubles'. It was also the year in which ownership of the local Public Analyst's laboratory passed to an English practice based in Chester. The partners of Ruddock & Sherratt assumed responsibility of providing the Public Analyst and Agricultural Analyst service in Northern Ireland.

Unlike the local forensic science laboratory, the Public Analyst's laboratory was never the target of a direct terrorist attack. Nevertheless its location on the fourth floor of a city centre building meant that it suffered the then unknown term 'collateral damage'. There was the troublesome, almost daily at some periods, evacuation because of 'bomb scares'. The staff often headed off to a pub in a 'safer' part of the city whence progressively more senior management would be dispatched to fetch them back once the alert was over, and sometimes they didn't come back either!

On many occasions bombs exploded in the vicinity, thankfully mainly when the building was evacuated, blasting in all the windows and damaging equipment. On several occasions, staff, including the Public Analysts, had narrow escapes when bombs went off in a wing of the building while the staff were still inside. The year 1992 was especially memorable for this. In that year, the forensic science laboratory was completely destroyed by a massive car bomb. The Public Analyst laboratory suffered in unrelated incidents, a car bomb in the adjacent street demolished a wall of the Public Analyst's office. The Public Analysts pressed on without an office wall, behind wooden shuttering until the wall was repaired, when it was promptly blown in again. Thankfully no staff were killed or injured and the laboratory continued to provide a service throughout, despite the disruption that often meant staff had difficulties even getting in to work. No samples or reports were ever lost or destroyed although some were written to the punctuation of nearby gunfire.

The laboratory in Belfast was not well equipped – it didn't make commercial sense to put expensive analytical kit next to 'bomb alley' but a close liaison developed between the resident Public Analyst, Michael Walker, and the Chemistry Department at the Queen's University of Belfast (QUB). This meant access to cutting-edge equipment. This was vital in solving a problem raised by an optician in Derry. One of his patients exhibited unusual eye dilation with no obvious physical cause. It reminded the optician of the effect of atropine and he called in the local Environmental Health Officer to investigate. Samples of herbal tea that the patient was in the habit of using came to the Public Analyst for analysis and rudimentary analysis at the Belfast laboratory suggested the presence of atropine. This was confirmed at QUB by more sophisticated tests under the direction of the Public Analyst. The herbal tea had been contaminated at harvest with belladonna (deadly nightshade) and the batch was recalled. Further support was provided by the Laboratory of the Government Chemist in London by Clive Shelton who later qualified as a Public Analyst. His recollections suggest that the patient, who drank comfrey tea, was experiencing rather strange dreams after her night-time drink (due to the atropine). He analysed the samples by quantitative microscopy (looking at individual fragments and counting them) to confirm the atropine chemical results.

Work continued despite the difficulties in Belfast and whilst the above demonstrates the tenacity and perseverance of Michael Walker, the stories below perhaps demonstrate the need for following a procedure, and what happens when it is not followed.

A YOUNG TOXICOLOGIST'S LESSON

During the early 1980s, a young novice chemist was finally, after several years of training, allowed to work in the toxicology department. This department deals with the analysis of specimens taken during post-mortem with a view to ascertaining if the deceased had taken poisons or drugs or therapeutic substances (medicines) and if these had contributed in some way to their death. As was the norm, a batch of samples of blood, liver, urine, and gastric contents relating to a deceased person was submitted to the laboratory along with a request from the pathologist for a toxicological screen. This always started with an assessment of the gastric contents. If a suicide was indicated, the gastric contents would often contain relatively large quantities of the substance of choice, assuming of course that the substance or substances were taken by mouth.

At first glance, to the novice chemist, it seemed an easy opportunity to discover the answer. Within the gastric contents he noticed a small fragment of foil with writing on one side. Examination showed that the letters on the foil were: 'ess'. No need to follow the routine procedure, a simple search of drug names, either trade or chemical names, should soon reveal the answer. This was made slightly more difficult by the fact that computer searching and web-based links were not freely available at that time. Armed with MIMS (the doctor's handbook), the BP (the pharmacist's handbook) and Clarke (the toxicologist's handbook) the search should be over in minutes; after all, 'ess' is an unusual combination of letters to find in a drug name, at least according to the young novice.

The search began and continued for some time, in fact for the rest of that day and into the following morning, but to no avail.

Then came the dawning realisation to the novice chemist there was no easy route to identification. Normal procedures then followed: the gastric contents were extracted into the four fractions – strong and weak acids, and neutral and basic substances (the normal place to find most drugs). A quick check of the cleaned-up extracts showed a likely substance in the gastric contents. The answer was now at hand. Having identified a possible substance it was time to move in for the final identification by checking the extracted residue against a standard of the suspected substance. Time to open the drugs cabinet to take out the standard material. A surprise was in store for the young analyst. The drug was wrapped in a plastic package with foil on the rear. No surprise so far, many therapeutic substances are wrapped in this manner. The surprise came when the wrapper was turned over to reveal the foil, which had written on it the word 'Press'; of course the magic 'ess' found on nearly every therapeutic substance which has been wrapped in this manner.

Lesson learned; procedures are written for a reason and should not be short-circuited. Not without very good reason and a lot more experience.

NEVER TOO OLD TO LEARN

Even experienced analysts get caught out some times. A food complaint was made by a consumer to Durham Environmental Health Department in the north east of England. Food complaints are made by members of the public who have purchased food from a retailer and which they feel is defective. Perhaps it was mouldy or contaminated in some manner or contained a foreign body. Under these circumstances the product may be submitted to Alan Richards, the Durham Public Analyst for inspection and analysis. The purpose of this is to establish the cause of the problem and to stop any recurrence.

The complainant met the Environmental Health Officer (EHO) and she duly recorded the details; when and where purchased, the nature of the problem, and where the food had been stored since purchase. This helped the EHO build the witness statement and chain of evidence so that, if necessary, a prosecution case could be built. Not all food complaints go to court but the details are best recorded as soon as the complainant makes the complaint, *i.e.* whilst the details are fresh in the mind. The complainant allegedly found a foreign body in a piece of chocolate cake. The foreign body was identified using low power microscopy as an adult *Ephestia ellutella* (cocoa moth, Figure 2.4).

Microscopical analysis and tests for phosphatase (a naturally occurring enzyme found in the digestive tract which is destroyed by heating/cooking) and trypsin (an enzyme used in the digestion of proteins which is also destroyed by heating) clearly showed that it had been cooked with the cake.

To help any investigation by the EHO, as was common practice at the time, the manufacturer was invited to view the evidence at the Public Analyst's laboratory by the EHO. The representative from the bakery turned up with the EHO and the evidence was laid out to see. It was often the case that photographs and measurement would be taken by the defendant to take back with him to assist the investigation at the factory. On this one (and only occasion) the defendant asked if this was the entire evidence available and he was told yes. Upon hearing this he reached forward, as if to examine the specimen, and then moved very quickly picking up the moth and eating the entire thing in front of everyone, thus destroying the evidence. In the absence of evidence no action was taken. The EHO

Figure 2.4 *Ephestia ellutella*, cocoa moth. Picture supplied by Alan Richards, Public Analyst.

and indeed everyone else learned a salutary lesson that samples must be sealed inside polythene bags before they are shown to traders.

Brian Dredge, a well-respected Public Analyst of many years standing, wrote to me when he heard this book was being assembled and his words are particularly pertinent here.

'The term "Public Analyst" comes from legislation and for years related to the official analysis of food and drugs. Prior to 1978 medicines dispensed and sold in pharmacies were regularly tested by the PA laboratory where I worked. I cannot remember anything unusual relating to an official sample but remember the trouble over a sample submitted by a patient who had been prescribed some barbiturate tablets and "knew" that the "latest ones were not right". I was asked to analyse them, but could find no barbiturate. The tablets were clearly impressed with the name of a well-known supplier and enquiries revealed that the tablets were from a batch of placebos that had been specially manufactured for the pharmacy of a local psychiatric hospital to treat the patient who was being taken off long term treatment, but had

become addicted. The doctor pleaded with my boss not to reveal our findings to the patient. My boss was a very thoughtful individual and probably thought long and hard about what to do. I don't know what the report of the analysis actually said (as an assistant analyst I was not privy to such information) but it was probably along the lines "The sample contained the prescribed amount of barbiturate" because I later heard that my boss had received a letter from the patient's solicitor threatening to sue him. I did not hear any more about the matter but I think that this illustrates that, on occasions, even to have the Wisdom of Solomon would not be enough when deciding how to report some results. Whatever they were analysing Public Analysts have always had to do more than simply apply a range of standard procedures in their investigations. Often the course of an investigation is dictated by initial observations and/or early analytical results. This is not always appreciated by lay clients, particularly those that wish to retain a very tight control over what the analyst should or should not do. On one occasion, I received from a school's cook an amount of frozen minced beef that had been supplied to her canteen, because she was "suspicious of it". We found that in addition to beef it also contained cured pig meat. Within a day or two of reporting our results I received a formal sample, ostensibly from the same supply, with instructions to analyse it "only for species, nitrite and nitrate". In fact it was from a different batch as there was no foreign species or curing salt (indicated by the nitrate/nitrite levels) present but as soon as the sample portion thawed it was clear that sample contained a lot (and I mean a lot) of extraneous water as the appropriate analysis showed'.

Are scientists natured or nurtured? This short essay doesn't seek to categorically answer the question. Certainly we learn to follow procedures because they are proven to work and give consistency, something which is dear to a scientist's heart. But as Brian demonstrates above, it is not simply about applying a test. The procedure on any forensic analysis will always involve detailed observations and hypothesis testing and deciding what procedures and tests are appropriate, a skill which is undoubtedly nurtured. Nevertheless people are attracted to science in the first place probably because of their character. Several people have joined the laboratories in which I have worked having thought science was the ideal career only to leave, citing it as boring and tedious. Viewers of TV forensic and police dramas be warned, it is not as glamorous as it may appear. Scientific questions are answered by careful and methodical analysis and research and knowledge, dull and tedious to some perhaps, but not to scientists. I suspect that scientists have a leaning towards the joys of science and then learn to appreciate the need for peer reviewed protocols which must be followed in order to achieve a fair test. Perhaps the argument against nurturing is made by the following interesting fact that in the history of Public Analysts there have been only three sons of Public Analysts who went on themselves to qualify. Clive Shelton, Michael Fogdon and Braxton Reynolds were certainly nurtured by their fathers. Clive recalls many school and university holidays being spent in his father's laboratory. When he finished his first degree he had the equivalent of around 5 years' on-the-job training to supplement his CV. Thus scientists are probably born and nurtured, but then, there are exceptions to every case.

REFERENCES

1. L. Cockcroft, *Working Life in London, Public Lectures and 'Operative Chemist'* Royal Society of Chemistry: http://www. rsc.org/Library/Collections/Historical/HistoricalCollection/ Accum/WorkingLifeinLondon.asp
2. http://www.bbc.co.uk/dna/h2g2/A492040
3. Taken from J.A. Secord, *Portraits of Science: Quick and Magical Shaper of Science*, and based on Pepper's *Boy's Playbook of Science*, Thoemmes Press, 2003 http://www.sciencemag.org/cgi/content/full/297/5587/1648
4. *Final Years and Scandal in the Library*: http://www.rsc.org/ Library/Collections/Historical/HistoricalCollection/Accum/ FinalYears.asp
5. Portsmouth Archives.
6. J.F. Liverseedge, *The Adulteration and Analysis of Foods and Drugs*, 1932, p 192.

CHAPTER THREE

Memoirs Clouded by the Mists of Time

This chapter consists of memoirs mostly provided by scientists working in Public Analyst laboratories which in some cases can no longer be completely substantiated; the memories have been 'fogged by the mists of time'. However, in most cases they have been corroborated by colleagues, but as they are memories from a while ago records no longer exist which can be used to verify every detail within the story, therefore they are as accurate as memory and the passage of time allow. In some cases the paper-work still exists and confirmation was straightforward. In all cases the stories help to demonstrate the wide variety of work undertaken by these protectors of public health.

THE OLD CHESTNUT: BEER TESTING

Perhaps the oldest anecdote is that Shakespeare's father was a tanner; a person employed to check the strength of beer. Tanners used to pour the ale onto a wooden bench to form a large but not deep puddle. They would don leather breeches, analogous to the trousers a fisherman may wear, except made from leather, and then sit in the puddle of beer waiting for it to evaporate. Once the evaporation was complete they would simply stand up, measuring whether or not their trousers would stick to the wooden bench. If it

Forensic Enforcement: The Role of the Public Analyst
By Glenn Taylor
© Glenn Taylor 2010
Published by the Royal Society of Chemistry, www.rsc.org

was difficult to stand up then this indicated that the beer was full of sugar, as the sugar would bond the trousers to the surface of the wooden bench. If the beer was full of sugar it was considered weak, low in alcohol, and probably adulterated; if it was easy to stand up then the beer was strong and not adulterated. There has been a great deal of discussion about the accuracy of this story and the reliability of the test. It is probably not as reliable as the other memoirs in this section of the book.

DRUNK AT THE WHEEL?

As long ago as 1930, driving under the influence of alcohol or drugs was prohibited in the UK under the Road Traffic Act. During the 1980s, if a driver was suspected of driving under the influence of alcohol then a sample of blood or urine was taken and divided into three portions. One portion was sent for analysis at a police forensic laboratory and the second was provided to the driver for independent analysis by a laboratory of his or her choice at his or her expense. The third sample was retained for reference in case of dispute. The drivers' samples were occasionally taken to the nearest Public Analyst laboratory where an independent test and report was provided. The legal limit for alcohol in blood was 80 mg of alcohol per 100 ml of blood. Whilst drivers are not permitted to drive with levels of alcohol above that limit it is still possible to be charged with driving or attempting to drive under the influence of drink even when the blood alcohol level is below 80 mg. These cases are rare but not unknown. In order to be certain that a driver was over the limit analysts allowed a tolerance of 6 mg/100 ml, effectively making the limit 86 mg/100 ml of blood. On most occasions the forensic scientists and the independent scientists obtained the same result. However, on one occasion the forensic scientists obtained a result of 87 mg/100 ml and the independent Public Analyst found a result of 86 mg/100 ml of blood. After deducting the tolerance there was a dispute about whether or not the driver was really under or over the limit.

This case was taken to court for a decision to be made. The Public Analyst, using the data relating to the accuracy of the method and several repeat analyses calculated that there was a one in four hundred chance that the driver was not over the limit at the time of the test. The court had to decide whether or not the driver was over the limit on the basis of 'beyond reasonable doubt'. Crown courts must be certain beyond reasonable doubt that a person is guilty before reaching a guilty verdict. In this case the judge decided that a one in four hundred chance that the driver was not over the limit could not be considered as beyond reasonable doubt and the driver was acquitted.

SCIENCE CAN PROVE NOTHING?

In 1990, the Hampshire Public Analyst laboratory received its largest ever environmental monitoring contract to assess the environmental impact of an industrial plant. Mathematical modelling was undertaken to determine the locations of likely pollution and therefore suitable places to monitor any possible impact. This mathematical modelling suggested the need to monitor six nearby locations and one additional site as a reference (an area which would remain unaffected by the proposed works). Monitoring commenced one year before firing up the plant and for one year afterwards (this is shorter than the norm but given the size and location it was agreed suitable). The monitoring looked at acid rain (sulfuric acid and hydrochloric acid), oxides of nitrogen, metals in air including lead, copper, cadmium, chromium, arsenic and nickel, and smoke.

On the agreed date the monitoring commenced and continued without a hitch. Occasionally analytical results showed pollution levels which were elevated, reflecting local weather conditions. For example, when the weather was cold and temperature inversions were noted, some minor pollution was seen in the immediate area. Likewise when winter fogs were seen, which have a tendency to hold pollution at ground level, traffic pollution was found in the area, particularly sites near to roads and towns. Temperature inversions are an interesting phenomenon; they can be noted on 'fresh', cold, sunny winter mornings when the temperature of the ground, and the air near to the ground, is colder than the temperature of the air at say 20 feet up. On these occasions the emissions from a chimney are then seen to travel back down towards the earth rather than skywards. This is due to the fact that the warm air travels towards a colder surface and the ground is colder than the air. On these cold days an increase in acid rain, smoke and some metals was noted – this was probably associated with traffic – and, occasionally, slightly elevated levels of mercury, possibly from the local crematorium, were also noted.

One of the locations was a nearby site of special scientific interest (SSSI), a forest. After consultation with English Nature amongst others, a hut in the forest with power was chosen as the secure area where equipment could be stored and used for the tests. The hut was adjacent to others and in a clearing in a slightly elevated position. The location coincided exactly with the results of mathematical modelling which showed that, especially at a time of temperature inversion, this would be a location for elevated levels of pollution, should any be noted.

The monitoring necessitated a weekly change of tubes, bottles and filters, *etc.*, which was undertaken every Tuesday. The analysis then took place back at the laboratory and was completed in time for the change-over the following Tuesday. All continued well until one day when elevated levels of chromium, arsenic and slightly elevated smoke levels were noted. Acid rain was not detected. This warranted a special visit by the project manager in an attempt to identify the cause. One conclusion was the burning of tanalised (treated) timber.

Whilst touring the area searching for clues the project manager was greeted by a ranger responsible for the forest. The conversation between the two went along the following lines:

> Forest Ranger: *Are you the guy who's doing the monitoring?*
> Project Manager: *Yes*
> Forest Ranger: *Why are you doing that here?*
> The project manager gave a long explanation about the mathematical modelling *etc.*
> The Forest Ranger continued: *You'll never find anything. That's barmy and a waste of money.*

The project manager proceeded to explain about the elevated results and where they may have come from. He said, '*Last Wednesday morning we noted an elevated level of smoke and chromium and arsenic which might suggest a bonfire which didn't burn for long and which didn't burn any plastic-containing materials*'. Whilst he was talking he noted that the colour in the cheeks of the forest ranger seemed to drain. As the project manager continued he noted a few pieces of timber in one of the huts in the clearing behind the ranger had been replaced, and upon returning his gaze towards the ranger noted that he was looking at the forest floor. The project manager's eyes turned down towards the ground where he noted the cold embers of a bonfire.

The forest ranger looked somewhat sheepish. Possibly he now realised the power of chemistry after all. He asked that the story be kept a secret, and until now it has been, and he probably hasn't had any more bonfires since that date.

HOW CLEVER IS THAT?

During the 1970s a sample was submitted by a member of the public to the local Public Analyst laboratory with a request for analysis. The member of the public had bought a bottle of distilled water from a local garage and alleged that it had damaged his iron. The complainant was asked to return to the lab in one week for his results and to pay for the tests when he collected the certificate of analysis. The sample consisted of an opaque, unlabelled, plastic bottle with a blue screw-capped lid. Subsequent analysis revealed that the contents of the bottle were in fact dilute sulfuric acid (approximately 33%), which is used for topping up car batteries and highly corrosive. It would certainly damage an iron.

As the garage was situated at the end of the road from the laboratory the Public Analyst decided to investigate during his lunchtime. Walking into the garage shop he asked the man behind the counter if he sold distilled water. He did. The man from the garage reached for a bottle from the shelf behind him and placed it on the counter. The bottle was opaque plastic with a blue screw-cap lid and unlabelled. The Public Analyst asked if he sold car battery acid. He did, again he turned around and took a bottle from a different shelf and placed it on the counter. The bottle was an opaque plastic with a blue screw-cap lid. Most important of all, it too was unlabelled.

The Public Analyst told him about the complaint and asked if he could have made a mistake. He said not – how could he – he filled the bottles himself? The PA suggested that he labelled the bottles in future and asked if anyone else worked in the garage. The garage owner said not and, after a few moments said, *'Wait a minute, I have recently taken on a young student studying law at the local polytechnic. He is about 6'1" has long blond hair and only works on a Tuesday'*. The garage owner then agreed that in future he would label the bottles.

The following week the complainant returned to the lab, paid for the analysis and asked for the report.

He was informed that the contents of the bottle were in fact sulfuric acid used for topping up car batteries and asked him if he was sold the product by a male who is around 6'1" tall, with long blond hair studying law at the local polytechnic and asked if he made his purchase on a Tuesday.

Without answering he took his report and sample, shook his head and muttered, *'These guys are good'* and left.

This was many years before DNA testing and shows the power of pure detective work. The Public Analyst is an expert practitioner, but in those days, science alone couldn't supply that much information.

INFESTED DATES WITHOUT ANY CASE LAW TO DEFINE INFESTED DATES

Rex Reeves qualified as a Public Analyst in 1974. He passed the MChemA (Masters degree in Chemical Analysis) after three long years of leaving the Hampshire lab at 3 p.m. to travel to the polytechnic of the South Bank, London, where he studied for two nights each week, returning home at midnight. He remembers 1974 as a memorable year for his profession; it was the centenary year of the Society of Analytical Chemistry and three candidates from Hampshire all passed their MChemA at their first attempt. Three in one year at their first attempt was probably a record.

To Rex the most important aspects of being a Public Analyst were not just that he would be an independent witness for the court, but also his honour, integrity and pride in being a member of the profession.

Approximately eleven years after qualifying he received a routine sample of dates from Southampton Port Health Authority. The dates were part of a large shipment valued at around £35,000. The samples were not packed ready to be sold as dessert dates. They arrived at the laboratory loose as part of a bulk consignment, so Rex had little information as to their final use; be it catering or to be packed as dessert dates. In order to provide a representative sample the Port Health Inspectors had submitted about six different samples to the laboratory.

Rex decided that these samples should be analysed for heavy metals, colours and preservatives and examined for bacteriological contamination. The results of tests did not show anything untoward.

Having given the samples some more consideration he decided to analyse them microscopically to look for filth and insect frass (excrement) in the areas around the stone. To his surprise approximately 5–7% of the fruit was affected, with one sample as high as 40% affected. This seemed very high to him as he had expected some infestation, but only around 2%. Now that he had encountered a problem, Rex couldn't find a quantitative legal standard to apply to the levels of infestation. When reporting results it is normal to consult guidance to aid the interpretation of the findings but there was no specific statutory instrument, guidance or Code of Practice or Food Standards Agency report to which he could apply the data. So Rex's final option was to consult case law to see what had been decided by the courts previously; he found nothing. The next step was to consult other experts such as MAFF (Ministry of Agriculture, Fisheries and Food) and the Tropical Products Institute (the trade advisers) and of course his boss who was new to the Hampshire lab at the time, John Fulstow. John was cautious about Rex's interpretation and the fact that other experts could not provide advice, other than MAFF who informed Rex that as the Public Analyst he was the expert and should reach his own conclusions. At the time Germany had a standard which was zero infestation. Should Rex apply that standard? He felt that this was too stringent because it is a natural product and some insect infestation was surely likely. At this stage Rex wondered about the wisdom of checking for insect infestation, perhaps he should have left it to chemical and bacteriological examination? After a lot of thinking, and considering the life cycle of dates, including pollination, infestation might feasibly occur, he decided on a standard of not greater than 5% infestation, feeling this reasonable for a natural product and defensible in the light of German standards.

Next he had to prepare his certificate of analysis. The regulations in place at the time stated *inter alia*: '*No person shall import into Great Britain from a third country any food intended for sale for human consumption (which is not an exempt product of animal origin) which is:*

- *Unfit: a danger to human health or hazardous to the health of humans*
- *Unsound; or*
- *Unwholesome'.*

There was little guidance on the last two categories and Rex didn't feel that the dates were a threat to human health, so therefore he opted for unwholesome. At the time this was probably the first food to be considered as unwholesome.

Certificate completed; all Rex had to do was sit back and await the court date. Whilst waiting he was informed that the consignment had apparently been rejected in both Canada and the USA. Rex wasn't informed why, but nevertheless he felt vindicated. The day was booked, the venue Southampton Magistrates Court, time to stand up and be counted and act as the expert witness. Before Rex attended court he was informed by senior managers at the port that the consignment was valuable and that there would be a major claim against the port if this case failed. Rex noted this and also was told by colleagues how clever the barrister and defence team were; they managed to goad him a few times during cross-examination. Rex was advised not to goad them back and not to 'take them on'. The verdict in court was that the dates were 'unwholesome'.

The defence team launched an appeal, which went as far as the Master of the Rolls; the case couldn't go much higher in the UK legal system. Rex nervously awaited the outcome. This case was referred for retrial at a magistrates court.

The second hearing was also at Southampton Magistrates Court. Through their defence team the importers announced that they would not use the dates as dessert dates but as ingredients. Rex thought that this was totally unacceptable. Infested the dates may be, but if they were minced and treated that would make it acceptable? Again the defence barristers were excellent and this time they had a full team of experts with them including, amongst others, the Tropical Products Institute (the people previously unable to guide Rex) and others. Rex suddenly realised how lonely it could feel as a Public Analyst. This time he wasn't simply there to act as an expert and give testimony; he was to sit with the legal team and advise them on the technical issues of the testimonies of others. He was definitely going to earn his money – an hour and a half to two hours in the witness box.

Rex was in the witness box for around two hours and cross examined by two different barristers and on more than one occasion provoked. *'You are a Public Analyst, what right do you have to set legal standards?'* asked one. Rex remembered the words 'don't goad them', advice from his legal team. He looked the barrister straight in the eye. *'I am not a lawyer, but I am trained in food law as part of my MChemA. The PA Regulations of 1957 might answer your question'*. Rex asked if the barrister was aware of these Regulations. As time went by, a matter of seconds that felt like minutes, Rex thought that the barrister wasn't aware at all. Rex had been admonished by colleagues in the profession and reminded that Public Analysts should be gentlemen and not engage in battles such as these, but he was young, the adrenalin was rising and he had been goaded. During this exchange Rex kept one eye on the JP and noticed a wry smile on his face; Rex thought that the JP also thought that the barrister was not aware of the 1957 Act. The JP asked Rex if he would enlighten the court. Rex quoted directly from the Regulations: *'Where there is no standard; legal, statute, or whatever, it is a duty of the Public Analyst to put a standard to the court for the court to consider'*. Enjoying the moment Rex went on to say that it was not for him to set a standard, merely to propose one and for the court to set the standard through case law. *'You might not accept my proposal but I have a legal right to propose a standard'*, said Rex. *'Standards of today often come from case law; standards proposed by my predecessors and agreed by court. But then as a lawyer you would know that!'* The barrister looked very cross and Rex thought; I am in for a hard time. He wasn't wrong.

Today most of our foods have applicable quantitative standards which can be applied but in this case there wasn't one so it was Rex's job to propose one.

'The barrister didn't appreciate what a Public Analyst's role was, so as a young upstart I wanted to teach him a lesson', Rex said during the interview for this story.

In court Rex went on to discuss insect life cycles, which I don't think impressed the JP too much, but fortunately he was given the chance to sum up his evidence. Rex stated that both the USA and Canada had rejected these dates and that the Germans had a zero standard, but if, however, after this evidence, you want the UK, through ports such as Southampton, to become the dumping ground of the world then that is your decision. Rex spoke from his heart and not his head on this occasion, and colleagues later told him that that was the way his testimony sounded, straight from the heart. At the end of his summing up there were some cheers from the public gallery which surprised Rex. The technical experts did not disagree and the science was never disputed. The Port Health Authority colleagues told Rex after the case was over: *'Forget the science – it wasn't that important in this case – it was down to your summary and how you presented the evidence'.*

Rex said at the end of the interview. *'I would have preferred to think it was the scientific evidence, but maybe it was the summary. I felt drained and euphoric at the end of this case and it will stay with me in my memory. It has also stayed in the memory of others at the port. I met one of the retired Port Health Authority Officers recently and he reminded me of the case. It seems it was also ingrained in his memory too'.*

TREATMENT FOR POISONING BY ANTIFREEZE

Ethylene glycol, the main component in antifreeze which is used to stop coolant systems in cars from freezing, in itself is not toxic, but it is metabolised to toxic chemicals in the body. This process normally leads to renal failure and death after consumption of only small quantities of antifreeze. The enzyme alcohol dehydrogenase is involved in the metabolism of ethylene glycol. Therefore, one effective treatment is to give the patient alcohol, which blocks the metabolism of ethylene glycol. However, high doses of alcohol also have toxic effects and therefore it is essential that if this treatment is to be used the patient must only be given sufficient alcohol to ensure that the alcohol dehydrogenase is occupied and that the patient doesn't suffer the toxic effects from too much alcohol. The treatment for a relatively small amount of ethylene glycol may last around 24 hours and needs very careful monitoring.

A Friday morning in 1986 and the phone rang with an unusual request from a local hospital. Chemists were informed that a garage mechanic had mistakenly drunk an unknown quantity of antifreeze from a plastic bottle, thinking it was a soft drink. The antifreeze mixture had apparently been repackaged into the soft-drinks bottle and left on the side in the garage. The mechanic, having realised his mistake, raised the alarm and was rushed to casualty at the local hospital and then to the renal unit where a quick-thinking medic knew of a treatment; keep the patient inebriated using alcohol and monitor his progress over the next twenty-four hours. The telephone request was to monitor the patient's blood alcohol levels over the next twenty-four hours. The target level was around 150 milligrams of alcohol in 100 millilitres of blood, about double the drink driving limit.

Fortunately, this was routine analysis at the laboratory and the team leader, Stuart Swain, was happy to oblige. Samples started arriving very late that afternoon and continued until Sunday. The lab was kept open to facilitate this work.

The good news is that the patient survived and probably advises that chemicals are never repackaged unless adequately labelled. Without the skills of a quick-thinking doctor and a chemist, this patient would not have lived.

DEATH BY ANTIFREEZE POISONING

A twenty-year-old man living alone in a flat in the south-west of England hadn't been seen by neighbours for a day or two and, although he tended to keep himself to himself, this caused concern. The young man was thought to be depressed and worried about money. The neighbour contacted the local police who went to investigate. When they entered the flat the young man was found dead. No suspicious circumstances were noted at the time and the death was referred to the local coroner who subsequently ordered a post-mortem examination. The coroner's staff then worked to gather further evidence.

The pathologist asked to undertake the post-mortem was Dr Adnan Al-Badri. He examined the body and recorded the following: the body was that of a young adult male with appearances consistent with the stated age. There were several linear scars consistent with self-harm on both forearms and a needle injection point in the right groin. The larynx, trachea and bronchi were congested; petechial haemorrhages were noted on the pleural surface of both lungs (small purple-coloured haemorrhages noted on the outer surfaces of the lungs). The central nervous system, respiratory system, cardio-vascular system, gastro-intestinal system, genito-urinary system, lympho-reticular system, endocrine system and musculo-skeletal systems were all thoroughly examined and nothing likely to result in death was noted. Finally histology (the study of tissues and organs using microscopical analysis) was undertaken. The heart was assessed with nothing likely to cause death noted. Petechial haemorrhages were noted on the pericardial (outside) surface of the heart. The liver showed a mild fatty change and congestion. The brain showed cerebral oedema (build-up of fluid) and the lungs were severely congested. In addition, focal pulmonary oedema was noted. This alone was insufficient to cause death.

Results of body fluid analyses (toxicology; a toxicological screen is undertaken in an attempt to identify substances to which a person has been subjected to or taken prior to death which may have contributed to the cause of death) showed that there was no indication of involvement of prescription drugs, nor of commonly abused drugs, including ethanol, in this death.

More evidence came through from the coroner's officer. Apparently the young man was last seen with a large glass full of brightly coloured liquid, which it was thought he had drunk.

This bothered Adnan. Young men do not simply die without reason; there were problems but no obvious cause. He looked further and noted that the renal tubules contained a large number of sheaf-like refractile crystals (see Figure 3.5). However, during post-mortem examination, no evidence was noted of renal stones.

It became clear to Adnan that there were crystals of calcium oxalate in the kidney. Calcium oxalate crystals are birefringent, *i.e.,* they refract light in two directions and shine when viewed under cross-polarised light, as seen in the image on the right hand side of Figure 3.5. This gave a clue – calcium oxalate crystals are formed during the metabolism of ethylene glycol. Now the story was beginning to formulate in Adnan's mind. If these were calcium oxalate crystals, and the brightly coloured liquid was antifreeze, this was probably the cause of death. An overdose of ethylene glycol can cause damage to the brain, lungs, liver and kidneys

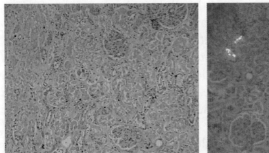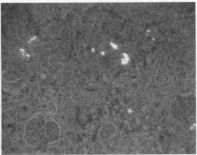

Figure 3.5 (Left) Section of kidney examined under non-polarised light. (Right) Section of kidney examined under polarised light. Images supplied by Dr A Al-Badri.

causing disturbances in the body's chemistry which may be severe enough to cause organ failure, and death.

Adnan concluded that the main finding of his autopsy was the presence of calcium oxalate crystals in the renal tubules of both kidneys. This was a sign of ethylene glycol (antifreeze) poisoning. Ethylene glycols are commonly used as antifreeze in motor cars and therefore can easily be obtained. Ethylene glycol is a sweet-tasting chemical that can be abused as a source of intoxication, drunk accidentally or used to commit suicide. Drinking in excess of 100–200 ml of antifreeze is almost certain to result in death unless specific and early treatment is given. The first symptoms of anti-freeze poisoning is similar to drinking excess alcohol and include nausea, vomiting and convulsions leading to coma and death within 24 hours. Detection of ethylene glycol in the blood depends on the time that has elapsed since ingestion. The blood sample that was sent for toxicological analysis was taken 14 days after death and hence was unlikely to test positive for this substance. The cause of death was given as ethylene glycol (antifreeze) poisoning.

TAKEAWAY ODOURS

Following repeated complaints from the occupants of a flat above a takeaway shop about odours from the shop, Glasgow Scientific Services were requested to help investigate. Environmental Health Officers undertook the first stage of the investigation taking samples of the air within the flat. The air samples (taken using special

42 — Chapter Three

sampling tubes) were sent to the Public Analyst lab of Gary Walker, and his team did a detailed analysis by gas–liquid chromatography and mass spectrometry. The scientists detected carbon monoxide and many food-related volatile organic compounds (odours), the profiles of which closely matched the atmosphere in the shop. Although a nuisance, this was not pursued because of the odours but because of the health issues related to carbon monoxide.

The scientists were asked to do a follow-up visit to the flat to investigate further. They undertook 'real-time' carbon monoxide and carbon dioxide monitoring to assess whether combustion gases from the gas cookers featured in the atmosphere in the upstairs flat.

The results (Figure 3.6) revealed significant levels of carbon monoxide in the flat, which was particularly alarming because the residents included a two-year-old child, and children are more susceptible to carbon monoxide poisoning. Exposure to levels such as these may have led to health effects such as headaches and nausea and worse if the exposure was prolonged. Furthermore, the pattern of peaks revealed that combustion gases, *i.e.* carbon dioxide and carbon monoxide, were present when the shop opened and disappeared when it closed.

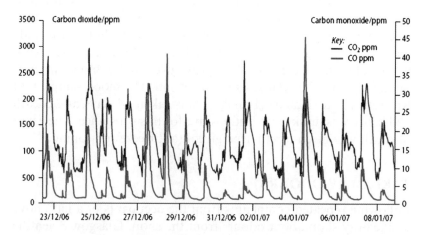

Figure 3.6 Carbon dioxide and carbon monoxide levels in a flat above a takeaway, recorded over a two-week period. Trace supplied by Glasgow Scientific Services.

Ventilation engineers assessed the takeaway's kitchen to find that the extraction system in use was inadequate. Furthermore, the pots used on the cookers were far too big, changing the burning characteristics of the gas-fired rings. They restricted the flow of oxygen and resulted in elevated levels of carbon monoxide. Carbon monoxide is produced when combustion is starved of oxygen.

What turned out to be an investigation based on nuisance (odours) had become far more serious with consequences for the health and safety of the people in the flat above and also the staff working in the kitchen. Environmental Health Officers served an enforcement notice on the shop requiring detailed remedial works and further monitoring. Once this was completed the problem was resolved. However, as a result of a number of other indoor air investigations in similar properties above takeaway shops, it became apparent that this was not an isolated case and odour complaints in these situations are now always accompanied by combustion gas assessments.

SIR, DO YOU HAVE ANYTHING TO DECLARE?

Since the 9/11 terrorist attack in the US, and subsequent suspected anthrax incidents, the UK has experienced a huge number of 'suspicious packages' and 'white powder' events. Invariably the police become involved in the investigation and in turn often rely on Public Analyst laboratories to determine the actual nature of the material. The analysis of such unknown materials presents a number of analytical challenges, which are exacerbated when the analysis is required to be performed in a suspected contaminated zone and it is therefore necessary to wear gas suits with breathing apparatus. While the majority of these incidents turn out to be hoaxes, the scientist must be able to show that the sample is free from biological material, radiological contaminations and chemical components that could be harmful to health.

A gentleman entered an up-market, Glasgow city centre hotel one evening pulling a suitcase, and tried to negotiate an acceptable price for a room for the night (Figure 3.7). Following much negotiation, he left the reception and was filmed on security cameras in a number of locations within the hotel before abandoning the suitcase in the hotel bar. The staff became concerned and called the police. Owing to the suspicious circumstances relating to this

Figure 3.7   Nothing suspicious... Image supplied by Glasgow Scientific Services.

incident, the fire service was called together with a scientist from Glasgow Scientific Service, the local Public Analyst laboratory.

Wearing suitable protective equipment, and after the police were satisfied that the suitcase did not contain an explosive device, the scientist inspected the case and its contents. Contained within the case was a 10-litre white plastic drum, spray-painted gold, wrapped in cling film, with three of the four upper edges cut away. A small amount of whisky-coloured, clear liquid could be seen within the drum.

The area was screened for radiation before a sample was removed to the mobile laboratory. Using a meter to measure α, β, γ and X-rays, the scientist established the absence of radioactive materials.

Only a few millilitres of liquid were available for testing and this was carefully removed to the mobile unit. Using a portable FTIR instrument, the scientist obtained an infrared spectrum and compared this to one in the instrument's library. There was little infrared activity and the predominant peaks were assigned to the presence of water. The pH of the liquid was near neutral. The adviser then did several chemical tests to rule out other common possibilities, which also proved negative. Finally the scientist checked the sample using high-powered microscopy with no further clues. The scientist took the remaining liquid back to the main laboratory where trace element analysis was carried out using inductively coupled plasma spectrometry and some further tests including ion chromatography. The liquid turned out to be predominantly water with a trace amount of detergent.

The mysterious gentleman was never found and the presence of this strange package contained within a suitcase remains a mystery.

'SUPERIOR' CONTAMINATED DRINKING WATER

During the mid 1990s a London restaurant refused to serve tap water, declaring that the bottled water offered was greatly superior. It could also charge for this superior water! The water was labelled in a similar manner to branded mineral waters, but microbiological examination and chemical analysis showed it to have a high bacterial count, together with yeasts, moulds and algae. It was high in sodium, low in hardness and high in nitrate levels. Alan Parker, at the time the Public Analyst to several London boroughs, deduced from the overall characteristics of the water that it was most probably tap water that had been softened and perhaps filtered before bottling.

Inspection of the restaurant by Environmental Health Officers revealed the water-bottling plant hidden in a corner of the cellar, comprising a water softener and carbon filter. Further investigation showed that the filter had not been changed for at least two years. The restaurant's 'superior' water was actually tap water, the wholesomeness of which had been significantly reduced by their intervention. Alan Parker remarked that *'the water sent to me was really in a poor condition'*.

SOMETIMES IT IS CRITICAL

During 2005 Nigel Wood, a food specialist and Trading Standards Officer, was contacted by a grandmother who had purchased some dark chocolate for her grandchild. The grandchild had an allergy to cow's milk. Children with allergies to cow's milk often have dark chocolate as an alternative to milk chocolate but they must take care to ensure that the chocolate does not contain any milk or milk solids. Apparently, after consuming some of the chocolate, the grandchild had an allergic reaction resulting in anaphylactic shock. This is an acute and sometime serious reaction to an allergen which can result in rashes, swelling of the throat leading to breathing difficulties, and a rapid lowering of blood pressure. In some circumstances these reactions can be fatal if left untreated.

The grandmother was, to say the least, mystified. She had carefully checked the label knowing that this was critical to her grandchild. The label did not state that the chocolate contained cow's milk, in fact it stated that the chocolate was dark so there should have been no problem. However, the child had still reacted badly to the chocolate. Nigel decided that further investigation was necessary and a member of his team submitted a sample to the Public Analyst to check the composition of the chocolate and checked the supply of the product. The results were returned from the laboratory: positive for cow's milk.

Further investigations showed the chocolate was manufactured in Belgium and imported through a supplier in Yorkshire. The importer had difficulty in understanding the problem as it turned out to be *'only a labelling error'*; the wrong labels had been used on a batch of chocolate products at the manufacturer. Eventually the importer was persuaded that this was a major problem and that Nigel could take action and would if necessary, under 'failure to withdraw' a product where the supplier knowingly allowed a product to continue to be sold in the knowledge that it does not comply with the legislation.

On hearing the story of the investigation the grandmother was delighted that the local Trading Standards department had taken the issue so seriously. The grandchild fortunately fully recovered from what was a frightening episode thanks to hospital treatment.

On recounting the story Nigel said: *'It might just be a technical labelling issue to some, but when you hear how pleased the grandmother was and the fact that the wrong label could cost someone their life, I remember why I do this job!'*

WHEN IS A TOMATO NOT A TOMATO?

On the 9th December 1993 a sample of a foreign material which was allegedly found in canned tomatoes was submitted to the Norfolk Public Analyst, Stephen Guffogg.

The foreign material was well travelled in that some of the specimen had been submitted to two other organisations for analysis before the Environmental Health Officer (EHO) submitted it to Stephen. The EHO said that he was dealing with a complaint of foreign matter and that two laboratories who had looked at it previously had reached different conclusions. He asked if Stephen would agree with one or the other or would he come up with yet another opinion. This was a difficult question as Stephen had not yet seen the reports nor the specimen. Stephen, ever the diplomat, tactfully suggested that he would have to analyse it first.

The specimen as submitted to Stephen was described by the EHO as follows: *'An irregular-shaped foreign body, the size of which is similar to a small bird's egg. Its dimensions are as follows: maximum length 22 mm; maximum width 15 mm; maximum thickness 13 mm. It has an off-white colour and on one side has a series of wavy but parallel markings of alternate dark and light colour. The other side presents a rougher more speckled appearance. Its general appearance and texture could be described as chalky'.*

Stephen's analysis started with a microscopical examination which showed the material to be crystalline and in one case a tomato seed was noted attached to the material. Stephen concluded that this, together with the staining on the outside surfaces of the material, was consistent with the material being found in canned tomatoes as alleged. Analysis by loss on ignition, chemical tests and high pressure liquid chromatography showed the material to be predominantly calcium citrate. Calcium and citric acid are naturally occurring in small quantities in tomatoes, but to form a crystalline agglomeration in the quantities noted would necessitate the addition of them to the tomatoes. Citric acid is occasionally added to tomatoes. Without further analysis, which he was not asked to undertake, Stephen could not identify the source of the foreign body.

The two organisations that previously analysed the specimen are both highly respected laboratories with international reputations and many years' experience of working in the food industry. One organisation concluded that the deposit comprised only natural tomato components, having apparently focussed their work on investigating if this could be the case. The other organisation used highly sophisticated techniques including electron microscopy and X-ray microanalysis which yielded the following elements: sodium, aluminium, silicon, phosphorous, sulfur, chlorine, potassium and calcium. They concluded that the foreign body was dampened and then dried agglomerated washing powder. These elements can be found in both washing powder and tomatoes as well as in other materials.

Stephen, as arbiter, proved the deposit in fact consisted mainly of calcium citrate (citric acid would not be detected using electron microscopy or X-ray microanalysis) and that the presence of entrained impurities probably caused the readings that resulted in erroneous interpretation and conclusions by the other laboratories.

Without further analysis Stephen could not conclude the source of the material so this doesn't answer the conundrum. It does, however, show that good old-fashioned chemical analysis combined with microscopy can, in the right hands, solve a problem, and that sophisticated technology alone does not guarantee the right answer without correct understanding and interpretation of test results.

MILK WITH THAT LITTLE SOMETHING EXTRA

A food complaint was received by a Welsh local authority Environmental Health Department in May 2006. The complainant had alleged that a foreign body was present in a plastic milk carton and was found when the complainant poured the milk on their morning breakfast cornflakes. They informed the local EHO that they washed the foreign body in the kitchen sink in an attempt to examine it and returned the remaining milk to the carton, giving both the milk and foreign body to the EHO.

The EHO decided to investigate further and submitted the specimen to the local Public Analyst, John Robinson. Upon receipt at the laboratory, inspection revealed an opened one-litre plastic semi-skimmed milk carton about a quarter filled with milk. No objects were visible in the milk through the side of the container. The contents were decanted and no objects were observed. It was noted that the milk possessed a 'faint sour odour'. Inspection of the related bag revealed the foreign body, which was subsequently identified as a small damp bat with an unfurled wingspan of 14 cm (see Figure 3.8). No milk was evident on the bat. The bat possessed a strong natural odour.

Given the evidence, it was not possible to conclude when or where the bat had entered the carton. The EHO could not rule out the bat finding its way into the milk carton post purchase at the complainant's property as it was found in a previously opened carton. However, there was a hypothesis that the bat had got in to the dairy through a faulty roof fan where the empty cartons were

Figure 3.8 The Public Analyst concluded that the bat was consistent with the species *Pipistrellus*. Images supplied by John Robinson, Public Analyst.

stored. These empty cartons were mainly wrapped in plastic film, but not all. One proposed scenario was that the bat crawled in to an empty carton to roost, which was then filled with the freshly pasteurised milk and sent on its way to the unsuspecting public. No droppings or signs of other bats were found at the dairy so consequently this could not be proved. However, the dairy was given an improvement order to seal the roof. The Public Analyst is pleased to report no further cases of bats in milk have been noted.

Perhaps the term 'enhanced milk' has another meaning now. Maybe this is enhanced milk with added protein plus whatever diseases the bat was carrying! Not a good way to protect bats, or the public.

MYSTERIOUS DUST

In 1995 a grey dust, appearing on windowsills at a nursing home, baffled the Environmental Health Department of Ceredigion County Council, the local authority. It resulted in an extensive report from an outside laboratory, citing very sophisticated testing. The laboratory issued a four-page report on the dust including the results of heating tests and Fourier transform infrared spectrometry (FTIR, sophisticated new technology at the time) which simply revealed that it was 'organic' in nature – something that the Pubic Analyst later thought was blindingly obvious – but the report revealed nothing specific and the cost to the council was £400.

Eventually a specimen was submitted to the local Public Analyst, Trevor Johnson, together with a request to identify the offending dust. As a trained Public Analyst of many years' experience, the first examination was microscopical, not perhaps a sophisticated and 'new technology', but in the right hands, very powerful. Immediately Trevor recognised the microscopical features of human hair fragments with interspersed skin tissue. This was compared the next day with his own face shavings from an electric shaver, which confirmed the nature of the deposit. It was identified as human beard hair! The Environmental Health Officer was most reluctant to agree with this finding; so he was invited to come down to the laboratory to view the specimens himself. This seemed to convince him and he undertook further enquiries which revealed that the dust was indeed human beard shavings. Apparently one of the residents regularly shaved other colleagues in the home and emptied the contents onto the windowsill of the home.

The Public Analyst charge for the examination and report was £40. From that day forward, the council have stopped using 'outside contractors' and have relied heavily on the good services of their Public Analyst department. Trevor states that: 'the fees have accelerated and modern techniques play a more prominent part in the extent of analysis that can be performed. However, the microscope is still the first port of call when examining pollution samples, complaint specimens, *etc.,* but is now usually used with additional adjuncts such as digital photographs and scanning electron microscope evaluation'.

CHAPTER FOUR

Some You Win and Some You Lose

The ultimate sanction in enforcement is inviting those thought to be guilty of a crime to court. It is not a sanction taken lightly; in fact very few Public Analysts attend court on a regular basis. Partly because their evidence is taken as read and agreed without the need for the author to attend court and partly because local authorities do all they can to avoid taking people to court, considering it far more preferable to work with a company to improve their produce rather than prosecute. In some of the cases in this book the Public Analysts were not aware that their evidence was used in court. My discussions with them as part of the research for this book was the first time they heard that the case was presented. Perhaps this demonstrates how their evidence is viewed by the legal profession, when it is used and agreed without the need for the author to attend. Nevertheless some Public Analysts are required to have their day in court.

LITTLE ABSHOTT MURDER[i]

On the 7th June 1996 in a peaceful farm at Little Abshott near Warsash, Hampshire, (Figure 4.9) a tractor driver was ploughing one of the fields ready to plant a crop of corn. Suddenly the plough behind the tractor struck what seemed at first thought to be a large

---

[i] Story and images courtesy of Hampshire Police.

Forensic Enforcement: The Role of the Public Analyst
By Glenn Taylor
© Glenn Taylor 2010
Published by the Royal Society of Chemistry, www.rsc.org

Figure 4.9   Little Abshott. Image courtesy of Hampshire Police.

clump of soil. The driver stepped out of the tractor to examine the
mass only to find what appeared to be a human arm. He immediately called the police.

When the police arrived it was clear that a human arm had
indeed been unearthed. It was noticed that around 40 feet away
along the plough line was a large swarm of flies. Further investigation was necessary, which included excavation of the deposition
site. A body was found face down in a sleeping bag wearing a
Pringle Sweatshirt (later identified as a copy), a yellow metal (gold-
like) ear stud in one ear and the hands were in front of the body
held between the thighs by a scarf. The skull was smashed, which
might have been due to the plough.

The body was sent for a full forensic post-mortem which included X-rays. As a result of the post-mortem the cause of death was
given as possibly due to the skull being struck by a blunt instrument. However, it could not be ruled out that death was as a result
of strangulation because of the damage caused to the neck and
skull by the plough blade. The body was badly decomposed and
consequently identification of the body was difficult, possibly a

white Caucasian man between 25 and 55 years old. (A decomposed body loses pigment so it can be more difficult to identify nationalities.) X-rays of the femur were taken to Louise Scheueran, an anatomist, who suggested that this was the bone of an Asian man. Fortunately the detective leading the team had recently returned from a debriefing of the Fred and Rosemary West investigation. He had heard of a forensic scientist, Richard Neave at Manchester University, who could reconstruct skulls using clay. This process took only ten days before a 'photofit' of the face was provided to Crimewatch, a TV programme designed to assist police in solving crimes (see Figure 4.10).

The scarf used to hold the hands together was sent for analysis of the knot used and to ascertain if any traces of materials were left which could help police. A forensic scientist at FSS Birmingham was able to identify the type of scarf as a Chunni, an Asian headscarf, commonly used by both males and females.

Two fingerprints taken from one hand which wasn't badly decomposed (because it was held between the thighs and therefore protected) were checked using the immigration database at Croydon but returned a negative result. They were also sent to New Scotland Yard and checked against the national database but to no avail.

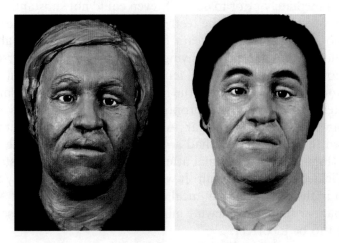

Figure 4.10 Images of the victim's face produced for Crimewatch by Richard Neave. Image courtesy of Hampshire Police.

A trichologist was consulted to look at the hair to try and ascertain if it was dyed and what the natural colour was. The hair was not dyed and not naturally black, *i.e.* not Oriental. The hair was tested for drug and poison residues and gave some positive results which were later discounted as false positive results probably due to decomposition.

A gastroenterologist examined the stomach contents and concluded that the last meal was a vegetarian chilli with kidney beans. Toxicology provided negative results for drugs or poisons in the blood and other organs. Again given the level of decomposition it was difficult to be conclusive about the toxicology.

Despite a possible name of the victim having been provided as a result of the Crimewatch appeal, together with the above results, a positive identification of the deceased was still not available. Even if there were murder suspects, it would have been virtually impossible to prosecute without an identification of the deceased. An odontologist was consulted to try to identify the victim from his teeth.

The odontologist, Steven Austin Jones, examined the teeth three days after the discovery of the body. The teeth were X-rayed and the molars were found to be fully developed. This indicated that the deceased was no younger than 18 years old. He noted that the teeth were slightly pink. This suggested that the victim may have been strangled, suffocated or asphyxiated or kept in wet/moist conditions immediately prior to death. Steven could not substantiate this theory but he could age the teeth and suggested that the deceased was no younger than 30 years of age. Steven was also able to conclude that Asian people living in Asia eat a more coarse diet and consequently their teeth tend to have less decay but more attrition, the converse being true for an Asian living in England, as they tend to eat a higher proportion of processed foods which contain higher levels of sugar. An examination of the teeth also showed that the deceased had a $\frac{3}{4}$ gold crown in one of the lower teeth (Figure 4.11). One final attempt to identify the pink staining also yielded a negative result. It was thought that the kidney beans found in the last vegetarian meal or beetle nuts (chewed by Asians in a manner similar to chewing gum) may have been the cause. This proved to be incorrect.

The advice of the scientists was that analysis of the crown would not achieve anything. The detective thought otherwise and decided

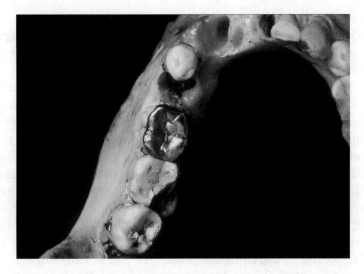

Figure 4.11 The victim's jaw showing a gold-capped tooth. Image courtesy of
Hampshire Police.

to have it analysed. The results were very unusual. Instead of
containing 60% gold the tooth contained 21% gold. This was an
excellent result. There was at the time only one manufacturer of
this type of metal composition who in turn supplied 179 dental
technicians throughout the UK. All the police had to do was talk to
all 179 dental technicians.

The slot on BBC's Crimewatch programme was a success. The
following morning after the programme on 3rd Sept., the name of
the deceased, Harjit Singh Luther, a 40-year-old Asian, was provided through Crimestoppers. This ultimately led to possible crime
scenes in Hampshire and one in Ilford in Essex. Fortunately a
dental technician, supplied by the manufacturer of the unusual
metal composition, operated in Ilford. At last a confirmed identification of the deceased seemed close.

Discussions with the dental technician in Ilford proved incon-
clusive. The dental records of the named person did not match the
deceased, a huge disappointment at the time. A revisit to the dentist
produced a surprising result, a different set of dental records and a
confession from the technician that he had defrauded the NHS by
charging for the more expensive gold crown. The revised dental
records were shown to Dr David Whittaker who confirmed that
they matched those of the deceased man.

Now, on the 10th Sept., warrants were issued to allow a detailed search of suspected crime scenes: a property in Shirley, one in Portswood, Southampton and another in Grange Road, Ilford.

Harjit was living with Manjit in Grange Road, Ilford. Harjit was already married and had three children by a lady who lived in India. He returned to India for three weeks and whilst he was gone, Manjit had an arranged marriage to Baljeet Rai. Baljeet was an illegal immigrant and needed to arrange a marriage to remain in the UK. It also transpired that Manjit was pregnant with Harjit's child. In late February 1996 Manjit decided to return to her former partner Harjit in Ilford. Worried that this would harm his chances of remaining in England and wanting Manjit to himself, Baljeet travelled to Ilford from his Southampton home.

The thorough search of the properties and a Vauxhall Cavalier car used by Baljeet and Manjit yielded the following: in Ilford, blood matching the deceased on the threshold leading from the landing to the main bedroom, blood on the floor next to the bed and also on the side of the bed, behind a radiator, on a chest of drawers and diluted blood on the polystyrene ceiling tiles. Considerable effort had been taken to clean the bedroom in order to hide the evidence. Also found at Grange Road was a vacuum cleaner with bloodstains, matching that of the deceased, on the end of the nozzle. On a cot later found to have been moved from Ilford to Southampton, were bloodstains.

It emerged that Baljeet Rai had returned to the Ilford property in the early hours of the morning to find Harjit in bed with his wife. His wife awoke and Baljeet Rai gestured for her to be quiet and get out of bed. She followed his instructions and picked up the baby and stood on the landing and observed the murder.

The choice of the burial site was due to the fact that Baljeet had worked at the farm and knew that 'spoil' from a nearby irrigation canal had been brought onto the farm and would offer a good hiding place. What he didn't know was that the farmer wanted to level the area to plant corn.

Subsequent enquiries yielded witnesses to the fact that the Cavalier was on the farm at or around the time of the burial.

On the 15th July 1998 Baljeet Rai was found guilty of murder and sentenced to life imprisonment.

The detective leading the hunt said that without the analysis of the gold crown, identification might not have been made; consequently a conviction would have been very difficult indeed.

This story does not emanate from a Public Analyst laboratory but from Hampshire Constabulary. Assaying gold and other metals is something that has routinely been undertaken in Public Analyst labs. This also demonstrates how far forensic science has travelled since the beginning and shows the variety of professions involved in a forensic case.

ADDED WATER IN CHICKEN BREAST PREPARATIONS – A LONG-TERM FRAUD?

The total UK market value for chicken and chicken preparations is just under £3 billion per annum. Chicken has grown in popularity as a low fat, high protein food or ingredient. The UK and European chicken producers are unable to produce all the UK's requirements for chicken, especially chicken breast. The catering and manufacturing sector import and use large quantities of chicken breast originating from Thailand (usually pre-cooked) and Brazil.

In the 1990s the Northern Ireland Public Analyst Michael Walker was intrigued to see a series of similar consumer complaints about chicken over a period of months. Each had the same characteristics – chicken breast, often purchased very cheaply from a 'white van' sale, was difficult to cook out properly, looked pink internally no matter how long it was cooked and was unusually 'solid' in texture when tasted – often to the point of unpalatability. The analysis indicated an unusually high content of inorganic phosphates and when raw chicken was available for testing the public analyst found an astonishingly high amount of extraneous water – often as much as 50% of the weight of the so-called 'chicken' was in fact added water. In Northern Ireland the problem was highlighted at a meeting of the NI Food Liaison Group and taken up by Gerry McCurdy, then a senior EHO with the Environment Dept. On inquiry, Public Analysts across the UK were coming up with similar findings. Accustomed to regulating by analysis the amount of added water in ham, for which a wet curing process was well known, Public Analysts were concerned that consumers and especially caterers were now unwittingly paying for tap water in a new and previously unadulterated product that was high on the shopping list for most households. Gerry McCurdy, as part of the transition team forming the newly created Food Standards Agency, brought the problem to Mark Woolfe, Head of Food Authenticity in the agency.

Chicken breast being low fat often has a dry texture when cooked. It had been found that succulence could be increased by adding water to compensate for water loss during cooking. A market opportunity was thus created and spawned a substantial industry, mainly based in the Netherlands, adding water to defrosted Brazilian chicken breast and refreezing the product before exporting to wholesaler distributors in the UK and Ireland. In 2008, the value of this market was estimated at around £200 million per year.

The usual process of adding water is by tumbling the chicken breast in a brine containing salts, sugars, possibly polyphosphates and milk proteins. This diffusion process is slow, and hence the amount of water that can be added to the chicken is limited. Change occurred when processors switched to multi-needle injection machines. Substantially larger amounts of water could be added in a matter of less than a minute. However, in order to hold added water of levels of 30–50%, much stronger water-retaining agents are required. The industry found that hydrolysed collagen proteins could hold 20–30 times their weight of water to withstand the freeze/thaw process and reduce cooking losses.

Mark Woolfe began investigating the UK market for wholesale chicken preparations (*i.e.* frozen chicken breast with added water and other ingredients, Figure 4.12) in 2000, as part of a survey on extraneous water in whole chicken and chicken parts. Concerns had been raised that the chicken breast preparations were not being labelled correctly, and Hull Trading Standards Department had already taken prosecutions. A survey of different brands was initiated in mid 2001 with the cooperation of 22 local authorities, who collected 68 samples of 10 kg cartons of frozen product. Analysis was carried out in 13 Public Analyst laboratories. Verification of the product and ingredient declarations had been facilitated by the following developments:

- A change to meat content declarations, which required 'chicken content %' rather than an added water declaration (in 5% steps)

- A revised nitrogen factor and an accurate threshold level for hydroxyproline (an amino acid found in collagen and connective tissue) for chicken breast, as a result of a Royal Society of Chemistry (RSC)/Analytical Methods Committee publication;[1] and
- Use of new and very sensitive PCR–DNA methods to detect species other than chicken.

The main outcome of this survey (FSIS No. 20/01 December 2001)[2] was to show that 20% of the samples used undeclared hydrolysed collagen proteins. 7% of the samples had under-declared the chicken content by more than 5%. Pork DNA was detected in two of the samples. There was, in addition, the misuse of poultry-meat marketing terms such as 'A-Grade' and 'halal' declarations (Figure 4.13). The Food Safety Authority in Ireland also carried out its own survey in May 2002, and found similar results, but with more positive DNA tests for both pork and beef.

Publication of the results by FSA and FSAI created much media interest and in 2002 BBC's Panorama devoted a full programme to the issue. The agency followed up the survey by requesting that 20 local authorities take formal samples, which resulted in formal action and one successful prosecution. The Dutch authorities took formal action in five plants involved with adding water in chicken

Dry breast fillet   Injected 70% breast fillet

Figure 4.12 Chicken breast preparations. Images supplied by Mark Woolfe.

Figure 4.13   Image supplied by Mark Woolfe.

breast and in some cases gave on the spot fines for breaches in labelling rules. Further surveys were undertaken as part of the Agency's Imported Food Sampling Programme, and the European Commission's Coordinated Control Programme. The labelling of meat products with added water was strengthened in the 2003 Meat Product Regulations, where the name should include not only added water but any ingredients not of the same animal origin as the food. There was also an audit of the FSA through a Food and Veterinary Office (FVO) visit to the UK and Netherlands and a Commission Position Paper on the use of water-retaining agents in chicken. In general, labelling of the cartons improved after these initiatives, although isolated cases of mislabelling were still found. Any subsequent DNA tests did not reveal the presence of either pork or beef DNA.

Apart from improvements to DNA detection using real time PCR, the Agency's programme of research was investigating the use of proteomics to detect offal tissues (liver, heart, kidney, *etc.*) in muscle, quantitative meat species determination, and the species origin of gelatine in non-meat based foods. The basis of these techniques is to isolate the proteins of interest, hydrolyse them (usually with trypsin or acid) and then identify peptide markers of the protein or species using high resolution tandem mass spectrometry. Just as specific sequences of DNA bases can uniquely identify different genes and therefore the species, so specific sequences of amino acids are biomarkers for specific proteins and species from which the proteins are derived.

The Panorama programme in 2002 had reported that the hydrolysed protein manufacturers altered the process to degrade the DNA so that it could not be amplified by PCR. This could never be proven by analysis until techniques were available to examine the hydrolysed proteins rather than analysing any DNA present. By 2008, methods were available which could achieve this, as well as other methods to look at whether mechanically recovered meat (MRM – meat recovered from the carcass after the prime cuts have been removed) or blood components had been used.

In 2008 there were two small UK producers of chicken breast preparations using multi-needle injection (Figure 4.14). As part of the evaluation of methods developed under the authenticity programme, the home authorities of these two companies collected the injection powders used in the production of different chicken preparations, and which were labelled as containing only hydrolysed chicken protein. There were two sets of powders from each processor, one used for products with 80% chicken content, and a combination of the two powders for products less than 80% chicken (usually 70% and 65% chicken).

The following analyses were carried out to elucidate the composition of the powders:

- Nitrogen and hydroxyproline were analysed by two Public Analysts
- The presence of blood plasma proteins by Nottingham Trent University
- Metabolomic analyses by Royal Holloway University (RHU) to detect the use of mechanically recovered meat and identify non-protein components
- Species DNA and amino acid analysis by Central Science Labs (now Fera)
- Proteomic analyses using a trypsin hydrolysis (University of York) and acid hydrolysis (RHU).

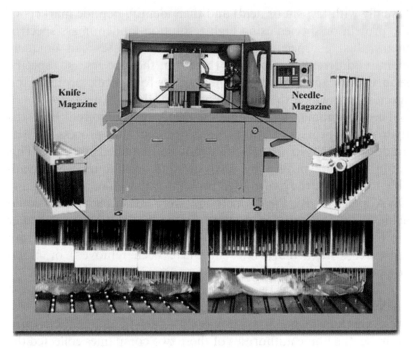

Figure 4.14 Multi-needle injection machine. Image supplied by Mark Woolfe.

The findings of this study (FSA 2009)[3] were:

- In all the powders the hydrolysed protein was derived from collagen, and no blood or MRM had been used. The same hydrolysed protein was used in both powders
- DNA analyses revealed strong chicken signals in all the powders, and one of the powders had a medium pork signal
- The two sets of powders had different hydrolysed proteins. Within each set, one powder appeared to be the pure protein, whereas the other powder was a mix of the protein with other components (maltodextrins, sodium citrate, sodium triphosphate, xanthan gum, *etc.*)
- In one set of powders, the trypsin-treated proteins, only bovine specific peptides were found, in the other set both bovine and porcine specific peptides were found. There were no avian specific peptides in any of the powders
- The acid-hydrolysed peptide markers confirmed the presence of bovine derived collagen in both sets, but although there was

a porcine marker found in the second set, this was not found to be specific just to pork but to avian species as well

- Similar bovine specific peptides were found in the exudates of a thawed Dutch chicken preparation with 70% chicken content
- Also a further sample of one of the proteins was collected by the Dutch authorities, who also confirmed the results using trypsin (Figure 4.15), which only found beef specific peptides and no avian specific peptides.

The scientists concluded that although both sets of powders were labelled as containing only hydrolysed chicken protein and gave positive chicken DNA signals, no avian specific peptides could be identified. The only peptides identified were of beef origin in one set of injection powders and beef and pork in the other set. These results were derived from one batch of powders from each UK plant, but similar beef peptides were found in a Dutch chicken sample and a powder sample collected 9 months later in the Netherlands. A later determination of the D-amino acid content indicated the proteins had been subjected to much more extensive

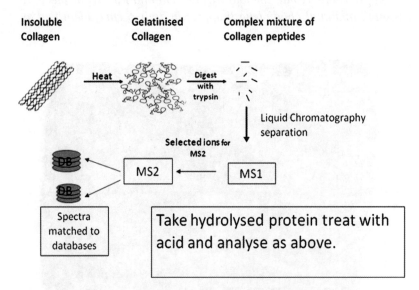

Figure 4.15 Collagen identification; Tryptic Shotgun approach. Image supplied by Mark Woolfe.

chemical degradation than gelatine. Hence the conclusion arrived at was that the powders only contained proteins derived either from beef collagen or beef and pork collagen, and chicken DNA was added to support the mislabelling of these powders.

This is a good example of where state of the art science eventually caught up with the mislabelling of these products. Unfortunately resolving this issue is proving difficult as both sets of powders are produced in other European member states, so access to the factories can only be through central and regional inspectorates, possibly with the assistance of the Commission.

DEAD MOUSE IN A MALT LOAF

A man bought a malt loaf from a supermarket in Ballymoney shortly before Christmas 2007. To his horror he discovered a dead mouse in the base of the loaf (Figure 4.16). He contacted the local council Environmental Health Department; Judith Freeburn investigated. She submitted the sample to the local Public Analyst Ron Enion on the 4th January 2008 and prepared the evidence.

Ron analysed the sample recording the following:

'Approximately half the loaf was present and was in one piece. A small mammal (body length approximately 6 cm), identified as

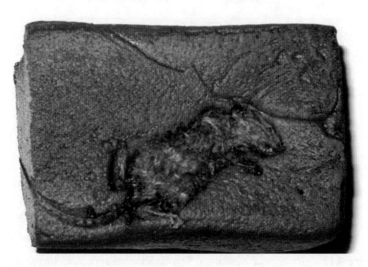

Figure 4.16   Dead mouse in a malt loaf. Images supplied by Ron Enion.

a mouse, was embedded in the base of the loaf. The mouse was embedded in, and was flush with the bottom surface of the loaf. When the mouse was removed it left a depression the same shape as the animal and did not destroy the integrity of the loaf.

The outer surface of the depression was continuous with the rest of the outer surface of the loaf; the crust surface had not been broken by the introduction of the animal. The colour of the crust surface within the depression was paler than that of the remainder of the outer surface of the loaf. These properties are consistent with the mouse having been embedded in the surface of the loaf before baking. The eyes of the mouse appeared "milky". An examination of a rear limb revealed that the skin and fur were easily separated from the underlying muscle, which was dry, pinkish and opaque. The muscle tissue gave a negative test for the presence of alkaline phosphatase. [Alkaline phosphatase is an enzyme found in body tissue which is inactivated by heat.]

The condition of the mouse and the manner in which it was embedded in the loaf, together with the absence of phosphatase, is consistent with it having been embedded in the loaf prior to baking'.

The defence lawyer argued that the mouse may have been placed in the baking tin as an act of sabotage. And that an 'onerous inspection' is held at the bakery every six weeks and that two field biologists attend each year. There are 131 bait stations in the premises at present.

The prosecution lawyer stated that the tins in which the malt loaves were baked were oiled the night before and then filled with dough; some time between the tins being oiled and filled the mouse must have got into the tin. *'In fairness to the defendant they have engaged pest control services who regularly inspect the premises and did so before this incident'*, she said.

North Antrim Magistrates Court heard the case and the judge fined the company £1000 plus costs for placing unsafe food on the market. In imposing the fine, the judge said he had considered public concern, but also the steps taken by the company to ensure proper hygiene.

MOUSE FOUND IN CHOCOLATE CONFECTIONERY

In April 1997 Ms Henriques had a chocolate bar bought for her from a kiosk in Piccadilly Circus, London. Having consumed most of the bar she bit into it for the last time and allegedly discovered something untoward. Ms Henriques informed the court that she was physically sick after eating a furry object.

Timothy Spencer, for Westminster Council, told Horseferry Road Magistrates Court, 'She partially unwrapped the chocolate bar, but noticed nothing unusual. She ate about three quarters of the bar and when she put the piece into her mouth she bit on something hard. She saw a black, furry object surrounded by caramel and nuts. She thought it was a peach stone. She showed it to a colleague who reported seeing thick black fur and red stuff. She said it was a mouse'. Ms Henriques immediately contacted the manufacturer who asked for a sample to be sent to them. She retained half of the object and supplied it to Westminster Council who submitted it to their Public Analyst, Alan Parker, along with a request to identify the foreign object without undertaking any further tests. Working in conjunction with the Natural History Museum Alan Parker concluded that the furry object was the head and shoulders of a mouse, which had a red colour deposited on its teeth.

The prosecution alleged that the rodent had entered the manufacturing process in the UK and therefore the manufacturer was negligent. Mr Spencer informed the court that the manufacturer had allegedly had a rodent problem at the time of manufacture (17th Feb. 1997) and that more than 80 mice were found at the time by pest control officers.

Mr Meyer, a rodent control officer for the manufacturer, informed the court that the company had previously had infestations, but that this mouse had, in his opinion, been brought into the manufacturer in nuts which were added to the confectionery. These nuts were imported from Turkey. The staining on the teeth of the mouse was thought to be due to a rodenticide used in Turkey, which was red, as opposed to the blue rodenticide used in the UK by the manufacturer.

The manufacturer was diligent in that it had regularly checked the importer and consequently Stipendiary Magistrate Tim Workman ruled that the manufacturer could not be held responsible as the mouse parts had been imported. He ruled that the manufacturer could not be found guilty of negligence under the Food Safety Act 1990. Westminster Council, who brought the prosecution, were not ordered to pay costs despite losing the case.

BLIND DRUNK MAY HAVE A NEW MEANING WITH THIS VODKA SEIZED BY CUSTOMS

John Robinson, the Public Analyst for Bridgend County Borough Council Trading Standards received a sample of '1806 Christoff Vodka 100% Pure Grain' (Figure 4.17) which had been sampled from a local public house for authenticity as part of routine market surveillance monitoring. Analysis revealed it to contain 0.75% methanol and 32.9% alcohol by volume (ABV) against a declared content of 37.5% ABV.

While this analysis was being undertaken customs officers, having received an independent 'tip-off', raided and closed an illegal

Figure 4.17 A bottle of '1806 Christoff Vodka 100% Pure Grain'. Image supplied by John Robinson.

bottling plant in Cardiff. Numbers of samples from this plant were submitted to the other Public Analyst in the area – the Public Analyst for Cardiff City Council (Cardiff Scientific Services). These included not only final bottled product but product still to be bottled. The methanol content of the bottled vodka was found to be at 0.66% with an alcohol content of 36.5% ABV whilst the methanol content of the liquid awaiting bottling ranged from 0.63% to 1.81% w/v.

The Food Standards Agency has stated that levels of methanol in vodka should not exceed 0.05% on safety grounds. Furthermore, there is a maximum prescribed level for methanol in vodka prescribed by the Spirit Drinks Regulations of 0.05%. This 'vodka' contained significantly higher methanol levels and has the potential to cause harm to any member of the public who consumes it. Symptoms could include blindness and kidney, liver and heart damage. While the concentration found was not sufficient to kill (a litre contains about a thirtieth of the minimum reported to be fatal), it could make you unwell and have lasting health effects.

It appears that unlike 'normal vodka' this drink was not a fermented and distilled spirit but processed industrial alcohol. The 'brains' behind the operation were not caught in the raid but 5000 filled bottles awaiting distribution and vats containing thousands of litres awaiting bottling were seized. This was a major illegal production plant. There is evidence that this product had been on sale since before Christmas as evidence came to light that a bottle was provided for a Christmas raffle. Not the Christmas present that most would hope for.

TOO MUCH HEALTHY FOOD CAN MAKE YOU SICK?

A Durham based consumer alleged that the muesli they had purchased from a local health food store tasted 'funny/disgusting after they had poured on the milk and started to eat it'. They complained to the local Environmental Health Department who carefully retained the whole sample and submitted it to the Public Analyst for analysis. The sample was analysed by Alan Richards, Public Analyst. Alan was informed that the 'health food' shop kept the muesli in a hessian sack and invited customers to help themselves using the scoop provided. This seemed to enhance the natural

Figure 4.18   (Left) Muesli with a little extra. (Right) The cat faeces. Images supplied by Alan Richards, Public Analyst.

qualities of the healthy food and probably reduced packaging and the burden on the environment.

Alan analysed the muesli and noted that the pieces of prune or date, yet to be confirmed by analysis, were covered in flour from the oats present. The shape, size and colour of the date/prune was as expected, but what was not expected was the discovery that it was not any kind of dried fruit but cat faeces coated in flour (see Figure 4.18).

During discussion of the results of the analysis Alan was informed that the reason for the presence of the foreign material was that the store was having problems with a mouse infestation which was eating the food so the owners brought in a cat to scare and/or kill the mice. It seems that the cat, however, could not tell the difference between the cat litter provided for the cat's personal use and the open sack of muesli. It was perhaps fortuitous that it had decided to defecate so that there was clear evidence of its presence. How long it may have been using the sack to urinate into no one knows – or probably wants to know! This is an occasion when the Public Analysts would not want to undertake a taste test!

FOOD ALLERGEN – PEANUT SABOTAGE, REGINA v. BENTLEY, NOTTINGHAM CROWN COURT

This was a unique case involving, so far as we are aware, the first food allergen incident involving deliberate tampering in a production facility. A maintenance worker accused of sabotaging a food factory by scattering it with peanuts was declared innocent after charges against him were dropped. Paul Bentley, who was allegedly reprimanded for having a 'girlie calendar' at work was said to have launched a revenge attack, Nottingham Crown Court heard during a trial which took place between 21st January and 6th February 2009. Peanuts had been scattered round a nut-free factory causing the plant to shut for 24 hours. Peanuts are a major food allergen and can cause potentially fatal anaphylactic shock to sufferers of the allergy. The closure and clean-up allegedly cost the company more than £1.2 million in lost production. The jury heard that a cleaner spotted 20 peanuts in an area of the factory where she had just seen Bentley. Over the next few hours more were found at different locations and at 4 p.m. managers decided to stop production.

The scientific evidence for the case was provided by the Food & Life Sciences Division of the Laboratory of the Government Chemist (LGC) led in court by Michael Walker. Food & Life Sciences Division were asked to become involved by their sister division LGC Forensics, the largest private sector forensic science organisation in the UK. Indu Patel of the food chemistry team at LGC Teddington tested a number of exhibits seized by police for evidence of peanut protein. Analysis and evidence recovery were undertaken in a dedicated restricted access allergen suite (see Figure 4.19). This is designed to guard against any possibility of cross contamination and includes checking the analyst's own diet to ensure that the food allergen had not been recently consumed.

Figure 4.19   (Left) Step-over barrier to restrict entry. (Right) Gowning-up, separate (colour coded) lab coats, and air-handling considerations. Images supplied by Government Chemist.

Figure 4.20 Swabbing procedures were developed by LGC especially for this case. Images supplied by Government Chemist.

Swabbing procedures were developed by LGC especially for this case (see Figure 4.20) and enzyme linked immuno-sorbant assay (ELISA) was used to detect any peanut protein which was present.

Peanut protein was recovered by swabbing the inside surfaces of the pockets of two exhibits of clothing allegedly belonging to the defendant which were used as part of the prosecution case (see Figure 4.21).

Partway through the trial Mr Bentley's defence requested experiments to assess the possibility of contact transfer of traces of peanut protein after handling peanuts with subsequent handling of garments and the effects of hand washing. They hypothesised that anyone coming into contact with peanuts and then subsequently handling the exhibits could transfer traces of peanut protein onto the exhibit thus contaminating them.

Instructed by the Crown, LGC set up an experiment to see if handling a peanut for 10 seconds transferred sufficient peanut protein to the fingers for it to be picked up from fabric even after 10 successive finger/fabric contacts.

The results and outcome of the tests indicated that even after brief contact with a peanut, peanut protein was readily transferred to clothing and easily detected by analysis and that only rigorous hand washing stopped the transfer. The defence argued that as it

Figure 4.21   (Left) Overalls. (Right) Tracksuit trousers. Images supplied by
Government Chemist.

was the same management team that found the peanuts who seized
the clothing, then there was a reasonable doubt that they contaminated the clothing.

The jury failed to reach a verdict having heard the evidence and
were discharged by the Judge. Consequently the defendant was
found not guilty. While it is the duty of every expert witness to
assist the court no matter which side has commissioned the work, it
is unusual for a single expert to give evidence for both the prosecution and defence in a criminal trial. The fact that all the scientific
work carried out in the case was conducted according to Government Chemist referee analysis protocols went a long way to
ensuring that the court was satisfied with the impartiality of the
evidence.[4]

REFERENCES

1. Analytical Methods Committee, Nitrogen Factors for Chicken
   Meat, *Analyst*, 2000, **125**, 1359–1366.
2. *Survey of Meat Content, Added Water and Hydrolysed Protein in
   Catering Chicken Breasts.* Food Standards Agency Food Survey
   Information Sheet No. 20/01, Dec 2001.

3. *Study into Injection Powders Used as Water Retaining Agents in Frozen Chicken Breast Products.* Food Standards Agency, June 2009; http://www.food.gov.uk/multimedia/pdfs/reportchi ckenstudy.pdf
4. Daily Mail on line, accessed 3 Jan 2009: http://www.dailymail. co. uk/news/article-1126351/1-2million-peanut-revenge-factory-worker-ticked-having-girlie-calendar.html#ixzz0bYiolRTa

CHAPTER FIVE

The Next Steps

Public Analysts are justifiably proud of their ability to apply their scientific knowledge to a practical situation and their ability to use classical techniques to identify fraud and adulteration. Over the 150 years since Public Analysts officially started science has developed, although the Public Analysts still use many traditional methods, such as microscopy, as well. This is not due to a reluctance to embrace change but simply that the standard tests and procedures have served them well and continue to do so. With the introduction of new technologies, however, has come the possibility of using these techniques for fraudulent gain, as seen for example, in the chicken meat fraud (Chapter 4). In addition, new initiatives such as 'organic', 'locally grown/regional food' will further challenge Public Analysts. To prove fraud in these areas requires the use of new technologies and techniques and consequently enforcement science requires the investment of time, research and money to keep up with new technology and potential fraudulent use of this new technology.

Science is developing at a rapid pace and many key developments are promising to have a significant impact, some of these are highlighted below.

Forensic Enforcement: The Role of the Public Analyst
By Glenn Taylor
© Glenn Taylor 2010
Published by the Royal Society of Chemistry, www.rsc.org

CAN A TIGER DETECT A GENETIC FINGERPRINT?

In the 1950s Francis Crick and James Watson furthered our knowledge about DNA (the basic building block of life which holds a cell's genetic code). Since then rapid improvements in technology have enabled the science to develop at an incredible pace. These developments have led to scientists being able to conclusively identify species from their DNA, for example to differentiate between halibut and haddock where substitution may have occurred. More recently DNA technology has been able to differentiate between individuals and has been successfully used in identifying those guilty of a crime when little other evidence exists. Since very little DNA may be available for analysis, it has been necessary for scientists to have the ability to multiply the amount available (amplify) and this has been achieved by using a technique known as real time polymerase chain reaction (PCR). PCR is an extremely sensitive test which only takes a few hours to complete and enables copies of DNA to be made allowing the detection of very low numbers of genes. These can then be quantified, enabling scientists to identify, for example, the amount of genetically modified material in a food.

After 9/11, the American government sanctioned research into novel detection techniques, particularly focusing on the detection of viruses and germs which were considered a possible 'white powder' threat. A $20 million spin-off from this anti-terrorist programme may fortuitously have had impacts on more areas of science than simply forensic science; now medicine and detection of food contamination are benefiting as well. Combination of PCR with a mass spectrometer (a mass spectrometer determines the mass of a molecule – some refer to it as the smallest set of scales in the world) in a technique sometimes known as PCR–MS or triangulation identification for the genetic evaluation of risk (TIGER).[1] This technique provides rapid reliable detection of genes associated with virus and bacteria and contamination of substances such as foods. This facilitates identification of a bacteria or virus in only a few hours which enables medical practitioners to treat a disease with the most effective treatment as soon as possible.

Recent trials have suggested that problems such as SARS or Group A streptococcal infections can be identified from a throat swab within twelve hours. If early indications prove to be correct then this technique could be very useful in reducing the spread of hospital infections and reduce the impact of food-borne illness from organisms such as *E. coli o157:H7*.

Fast identification is essential in restricting outbreaks. TIGER's faster detection could lead to faster conclusions regarding the source, thereby reducing the number of infections and saving lives. Additionally a key benefit will be the ability to detect and identify smaller quantities of the bacteria, amounts which currently may be considered too small to detect. Rapid genetic fingerprinting of viruses and bacteria will enable better epidemiological tracking of outbreaks. Perhaps this technique could have aided the tracing of the recent outbreak of Swine Flu (H5N1) from Mexico and would facilitate earlier detection of variations as new organisms are produced.

DOES TRAVEL DO MORE THAN SIMPLY BROADEN ONE'S HORIZONS?

Many say that travel broadens one's horizons, but after new research it may leave more of an impression than previously anticipated. '*I know where you have been, what you ate and who you were with*' was the opening gambit by Rebeca Santamaria-Fernandez in the 2009 Government Chemist dissemination seminar. She laughed and then said, '*Well maybe not just yet*'. Her work and the work of others like her suggests that in the near future it will be possible to tell much more from hair samples measuring only a few centimetres. Hair is not traditionally considered the best specimen for forensic work as it can easily be affected by environmental factors such as pollutants. However, Rebeca's work is proving to be very promising and unsurprisingly it is subject to a great deal of interest in the forensic world as, for example, it could completely destroy a terrorist's alibi.

Her ambition is to demonstrate if and where in the world people have travelled and possibly show the impact of the food they have eaten, all from one single hair strand which measures less than 5 cm long. In order to carry out this study, Rebeca and her co-workers focused on the most abundant sulfur isotopes in hair keratin – sulfur-32 (^{32}S), which accounts for about 95% of total sulfur, and sulfur-34 (^{34}S), which accounts for approximately 4%. Sulfur levels in hair are around one part per million (one droplet diluted in a car tank full of fuel).

The method involves the coupling of a laser system to a mass spectrometer. The laser makes contact with the selected fraction of the hair, eliminates surface contamination (coming from hair products) and generates an aerosol containing sulfur. This is later ionised within a plasma mass spectrometer (multicollector inductively coupled plasma mass spectrometer, MC–ICP–MS). The resulting measurement provides the exact proportions of the sulfur isotopes in the hair. Those proportions would vary slightly from one individual to another, and also within one hair strand, and those variations seem to correlate with changes in geographic location and/or diet changes.[2]

Since hair grows at a rate of approximately 1.25 cm per month the data taken from 5 cm of hair can be used to 'demonstrate' its owner's activities over the last few months before the hair sample was taken. Using differences in isotope ratios found in the hair after travelling, Rebeca and her fellow scientists are able to provide information as to whether a subject has travelled within the past few months. This was recently demonstrated by subjecting three individuals, two UK residents and one traveller, to isotope ratio analysis of $^{34}S/^{32}S$ isotope ratio in keratin (a protein which is the major component of skin, nails and hair) along the length of a 4.3 cm long strand of their hair. Difference in sulfur isotope ratios are plotted as $\delta\,^{34}S$ values where:

$$\delta^{34}S(‰) = [(^{34}S/^{32}S_{Sample}/^{34}S/^{32}S_{Reference}) - 1] \times 1000$$

The results are shown in Figure 5.22. Slight variations were noted in the two UK residents (around 0.1%, possibly due to differences in diet) whereas the variation noted in the traveller – a person who had just returned from four months of travelling in Croatia, Austria and Australia, with stays of over one month in each and has since then been living in the UK for the past couple of months – was greater than 0.5%. Rebeca concludes that '*measurement of δ^{34} sulfur variations has the potential to be an indicator of geographical origin and recent movements and could be used in combination with measurements in water and foodstuffs from different geographical locations to provide important information and geographical studies. Further work will involve the collection of*

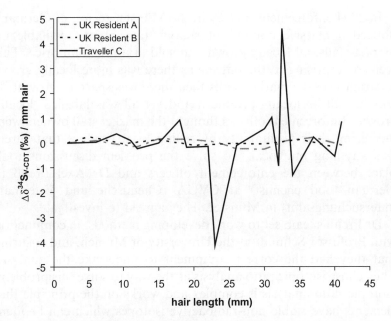

Figure 5.22 Sulfur isotope ratios expressed as $\delta^{34}S$ measured over 4.3 cm of hair are plotted for three individuals. (Reproduced with permission from LGC).

samples from individuals of both different geographical origin and different diet to evaluate a potential correlation between provenance/ diet and isotopic variations in hair'.

GERMAN BRAINS LEAD THE WAY ON A WORLDWIDE 'EASY' FOOD-FRAUD

The Germans, like their other European counterparts, are becoming increasingly aware of 'food miles'; consumers want to reduce their personal impact on the environment by buying locally grown and produced food. Consequently anything locally grown attracts a higher price. Is it really locally grown? How can the consumer tell? Is labelling in shops and on market stalls correct? Surely this is an easy fraud, just label the food as local and charge a premium, especially on market stalls where food often isn't sold with a label or the labels on the stalls are generic. Until recently forensic scientists hadn't developed methods which enable them to identify locally grown produce.

In 2004 enforcement officers in the Münster region of Germany noticed a massive amount of spargal (asparagus) available on market stalls; all locally grown and sold at a premium price. This seemed surprising to the officers as there was more locally grown asparagus on the market stalls then there was space to grow it. It couldn't all be locally produced and yet all was labelled 'locally grown'. Enforcement officers thought the market stall owners were increasing their profitability by selling local produce that surely was anything but local. To solve the problem discussions took place between the enforcement officers and Dr Axel Preuß (a German food chemist) at CVUA (Chemische und Veterinär-untersuchungsamt) in Münster. They agreed to investigate.

Dr Preuß's team set to work developing methods, in conjunction with Professor Schmidt at the University of Munich, and ensuring that they had the correct equipment to undertake the analysis. They chose isotopic ratio analysis as the way to solve this problem. Isotopic ratio analysis is complex and works on the principle that elements have stable non-radioactive isotopes which can be identified and quantified to allow the ratio of isotopes to be calculated. If this ratio is compared to information about the levels of isotopes in a region then the region where the food was grown can be ascertained. The origin of water used to grow crops or provide drinks for animals can be identified by isotopes of oxygen. Thus the water trapped in a plant can be used to identify the region from which that plant had grown. Combine this with isotopes of carbon, nitrogen and sulfur and it is possible to identify the region from the soil. The investment needed was around £250,000 for equipment alone and then in addition there were the costs of development time and training of staff, but they realised that this technique would be very useful in identifying other fraud as well.

Dr Preuß said, *'We chose to analyse δ^{15}nitrogen, δ^{13}carbon, δ^{34}sulfur (from the protein of the asparagus) and δ^{18}oxygen (from the vapour/head space above the asparagus juice) and determined all four isotope ratios to be on the safe side. In the first year (2006) we found more than 50% of the asparagus samples were not from the home region as claimed; they were from France, Spain and other regions in Germany. The fraud rate reduced to less than 5% in 2007 and again in 2008. The traders learned obviously that now we can prove the fraud'.*

Having perfected the technique the team were able to turn their attention to high-quality sparkling German wine. Under German law real Champagne or 'Sekt method Champenoise' (manufactured by the Champagne method) as well as normal Sekt ('*Sekt*' is the name for sparkling wine produced in Germany) contain only carbon dioxide which was produced by the yeast during the fermentation process without the addition by injection of technical food grade carbon dioxide. In Champagne the carbon dioxide was produced directly in the bottles while in normal Sekt it was produced in (stainless steel) vats. But in both cases no technical carbon dioxide is added. In ordinary sparkling wine (in German '*Schaumwein*') the carbon dioxide comes more or less out of steel bottles from the technical gas industry. This carbon dioxide is of food grade quality, but nevertheless it is a fraud to add it to high quality Sekt or Champagne.

> Dr Preuß said, '*In the case of wine we examined the quality of carbon dioxide (CO_2) through the δ^{13}carbon and δ^{18}oxygen isotopes. With the isotope technique we are able to show if in the bottle is original carbon dioxide from the wine alcohol or [if] technical carbon dioxide was added. In a few cases we were successful with this new technique and demonstrated the fraud*'.

As a result of this development work and sharing their expertise, other laboratories are now learning how to identify locally grown products. This could become a very powerful tool in the enforcer's armoury to help detect food-fraud around the world.

YOU ARE WHAT YOU EAT

You are what you eat may be more accurate than we might imagine. Isotopes in the food and consumed when we eat are absorbed into our bones and teeth. These isotopes can then be identified by the techniques discussed above. This enables scientists to identify our geographical origin. This technique was used in 2001 when the torso of a Nigerian boy was found in the river Thames and, without information to identify the body, police gave him the name of 'Adam'. Dubbed by the press as the 'Torso in the Thames' it was suggested that he had been killed in a muti (ritualistic) murder. A post-mortem was carried out and from the food in his stomach and the type of pollen in his lungs it was established that he had not been in the UK for long; only a few days. Police needed to identify where he had come from. Adam's bones were analysed using isotope ratios and the results compared to geographical databases of isotope ratios and the data confirmed that he had lived in the Yoruban Plateau in Nigeria.

Whether or not techniques such as these are fully adopted by enforcement scientists remains to be seen, however, metaphorically, the Public Analyst profession has reached a crossroads. The business future is uncertain and consequently there is a lack of investment/coordination, but science is improving rapidly necessitating significant investment in new equipment and training of staff in new technology. Something has to change.

REFERENCES

1. S. A. Hofstadler, *et al.*, 'TIGER', *Int. J. Mass Spec.*, 2005, **242**(1), 23–41.
2. R. Santamaria-Fernandez, J. G. Martínez-Sierra, J. M. Marchante-Gayón, J. I. García-Alonso and R. Hearn, 'Measurement of longitudinal sulfur isotopic variations by laser ablation MC–ICP–MS in single human hair strands', *Anal. Bioanal. Chem.*, 2009, **394**, 225–233.

CHAPTER SIX

The Final Chapter?

Is this the final chapter for Public Analysts after 150 years of fighting to improve the quality of food and stop food-fraud? Increasing costs of science coupled with a decreasing amount spent on enforcement, the desire of Government and others to reduce red-tape, and a perception that food offers a low risk to health, has led to many requests for a less regulated food industry. This consequently has led to a reduced Public Analyst service. Enforcement science has, however, never been better and interest in food and diet has never been higher. Science in general and enforcement science in particular have changed dramatically over the last thirty years:

- New technologies have been introduced which in the right hands can detect ever-decreasing amounts of harmful compounds in our food
- New techniques such as DNA identification and isotope analysis have together made fraud in food much more easily identifiable and will increasingly do so as these techniques improve; and
- Management of science.

European Law[1] has led to changes in the way in which we manage science and demands:

- Full accreditation (external 'official approval' of the work undertaken by Public Analysts)

Forensic Enforcement: The Role of the Public Analyst
By Glenn Taylor
© Glenn Taylor 2010
Published by the Royal Society of Chemistry, www.rsc.org

- External proficiency trials (proof of the competence of the analyst and laboratory); and
- Validated methods of analysis (proof that their methods work at the necessary concentration in the materials they are testing).

This legislation has resulted in increased certainty around the work of scientists. The example of blood alcohol levels (in Chapter Three) where an allowance for uncertainty was deducted before providing evidence in court was rare thirty years ago, now it is the 'norm'. Public Analysts have, ironically, never lacked confidence, but they now can demonstrate why they are confident of their analysis. Scientific enforcement has never been better, more highly regulated and consequently costly. Added to this, UK law requires that food samples which may be used in a prosecution be divided into three portions: one for the Public Analyst; one for the food company to have tested independently; and one, should there be a dispute, for the Government Chemist to test to provide the 'definitive result'.

In the late 19th century Public Analysts somewhat resented the Government Chemist 'looking over their shoulders'. Their objections centred on the competence of the chemists of the Excise Laboratory at Somerset House, the forerunners of the modern Government Chemist. A typical remark is that of Hassall: '*they* [the Somerset House chemists] *have afforded few public proofs, so far as we are aware, of their competence for the duty imposed upon them*'.[2] Further details of the uneasy early relationship between Public Analysts are noted in a history of the Government Chemist's laboratory.[3] However, the history notes that by the early 20th century relationships were cordial and so it has continued to the present day. The Government Chemist (now a private company rather than the former government department) cooperates with the Association of Public Analysts' (APA) Educational Trust in running three annual seminars and an intensive annual week-long residential course to update Public Analysts and their staff and is open to all stakeholders. The Government Chemist is generally represented at the annual APA conference, often by a nominated officer who gives a review of referee cases. Cooperation continues to build with joint research work and an open invitation for Public Analysts to avail themselves of the advice from Laboratory of the

Table 6.1 Comparison of results from LGC and PA of analysis of disputed samples

Year	2007–8	2008–9
No. samples referred to referee	21	25
No. samples where LGC agreed with the findings of the PA	18	23
%	84.2	90.9

Government Chemist scientists in cutting-edge analytical technology such as DNA and isotopic techniques in particular.

Over the last two years Government Chemist (LGC) released information relating to disputed samples which demonstrates how often its results are similar to those found by Public Analysts (PA), see Table 6.1.

The stakes are high for any expert witness, particularly Public Analysts, and indeed any forensic scientists. In his essay in 1923 on the *Expert Witness*, C.A. Mitchell[4] wrote:

'Mistakes in the observation of facts which are the premises for the conclusion are naturally much more vital than mistaken deductions, which cannot always be disproved, even when they are false, and that is one reason why the world judges more harshly the mistakes of observation, than those of deduction. It was a frequent lament of Dr Campbell Brown, a former Public Analyst of Liverpool, that a mistake made by an analyst was usually regarded as unpardonable. "A Doctor", he said once, "makes a mistake and buries it. A clergyman makes a mistake, and it is only discovered in the next world. A lawyer makes a mistake, and is paid for it as highly as if he had not. But if an analyst makes a mistake he is condemned".

.... This popular view arises from the remains of the old idea that the chemist practises the black art. He is supposed to perform some simple, though mysterious, magic on a thing, and presto, he knows all about it. If he makes a mistake, that indicates that his magic is bad; he is not a true magician, but a false quack.

That this is always the fate of a chemist who is at fault in his facts is disproved, however, by Campbell Brown himself, by the following instance within his experience:

"An analyst who had expressed a certain opinion, upon which action had been taken, found before the hearing of the trial that, misled by a defective instrument, he had made a mistake. Although the solicitor on his side strongly urged him not to mention the mistake, but to allow the accused person to be acquitted as the result of the conflict of scientific evidence, he felt that he must be absolutely candid, no matter what the consequences might be to himself from his side. At the trial he at once admitted the mistake, and the case was dismissed with heavy costs against the prosecution. But in spite of this, the ultimate result to the chemist was that for years afterwards his evidence was accepted as that of a man who was absolutely trustworthy".'

Improving science doesn't come cheap, the cost of accreditation alone is estimated at 15% of the total budget/turnover of the laboratory. This is at a time when the prices charged for analysis by Public Analysts are being driven down by competitive tendering and spending on enforcement by local authorities has been substantially reduced. The situation is, to say the least, perilous for Public Analysts. PA laboratories have been closing at a rate of one per annum over the last ten or so years and now only 17 UK labs remain. The most successful commercial analytical labs in the past 20 years have been labs other than PA labs. Public Analysts are good scientists but not trained as business managers; lower income brought about by dropping prices charged when costs are rising led to reduced viability. To balance the books most Public Analysts cut investment in training and new technology and saw other Public Analysts as competitors and refused to collaborate. This has led to a fragmentation of Public Analysts who consequently refuse to share methods *etc.* in the hope of maintaining competitive advantage. Frankly, Public Analysts do not have an answer to this conundrum; many have had to supplement meagre enforcement incomes with other, more profitable work in order to keep open. Combine with this the risks associated with providing a wrong result and the associated damages which can run to tens of thousands of pounds and it's becoming a risk too far for some local authorities who own their own Public Analyst laboratory. But these authorities are shoring up the enforcement system by keeping their Public Analyst laboratories open. Consequently many doubt the future of enforcement science as it is configured today; has the death knell of the Public Analyst sounded?

Food-fraud and adulteration were local issues in the 1860s. Local manufacturers such as bakers and brewers adulterated their products, local retailers followed the example to increase profit-ability, and local chemists fought the fight. The Adulteration of Food Act 1860 was not judged to be a success because too many local authorities failed to act by appointing Public Analysts and some failed to allocate any resources. Several attempts have been made to improve the system. One revision of the Act dictated that all local authorities had to appoint Public Analysts. This they did, but still they failed to allocate any resources.

Over the next hundred years the problem developed and became more regional than local, but the imbalance in the way fraud is fought amongst local authorities increased as some local autho-rities left it to others because the legislation didn't decree what they must achieve. The problem is that some local authorities have never seen it as a priority preferring to let others take the mantle in the fight against fraud.

Now the problem is worldwide, as demonstrated by the Chinese melamine debacle where a chemical was added to foods to defraud the consumer by foiling attempts by enforcement scientists to detect lower levels of protein in food, thus allowing the sale of fake products. Dr Andrew Wadge, the Chief Scientist for the Food Standards Agency, posted in Food-Fraud on the 24th Sept 2008:

'The scale of the problem in China caused by adulteration of baby milk with melamine is appalling. I struggle to comprehend how people can knowingly put the lives of so many babies at risk by adding this substance to milk. I spoke about the regulation of the food industry on the Today programme and BBC Radio Scotland yesterday morning. Of course it would be naive to assume that criminals will not attempt to target the food industry here and we know, through attempts to sell counterfeit vodka containing dangerous levels of methanol, that they will. However, there are, fortunately, big differences between China and the UK and in many ways what is currently happening in China mirrors what we saw in the UK 150 years ago, when adulteration and poisoning from food was commonplace'.

What happens in China today arrives in Europe tomorrow through import at the request of a more cosmopolitan population who demand a wider selection and cheaper produce.

Yet the success over the last 150 years has led many holders of the local authority purse to believe that the fight has been won and suggest that resources should be allocated elsewhere in the authorities' complex list of responsibilities. Rogers, Hampton and others have suggested that food composition is not a priority for enforcement officers. Combine this with the requirement for a lower burden for industry and this has led to many local authorities understandably questioning their commitment and spend on food law enforcement. At the Society of Chief Trading Standards Officers meeting in Solihull on 28th November 2007 the society concluded: 'There is a general concern amongst Chief Trading Standards Officers about this position but a recognition that at a local level food enforcement does not have a high enough priority to achieve additional funding from their own local authority, or a greater share of existing funding that goes to the local service. The Society's view is that the reduction over a period of years in spending on food enforcement is regrettable but in the light of local and national priorities, and a backdrop of tighter budgets, very understandable. However, the Trading Standards profession should be aware of the ramifications of these local decisions and should consult with national bodies to ensure they also appreciate the situation and to determine whether jointly any ideas can be formulated to improve the national position'. Did that sound the death knell?

In accordance with the Beijing declaration signed by over 50 countries, all signatory countries have agreed to develop comprehensive programmes of monitoring food safety on behalf of their citizens; systems are absent in many countries, and where they are in place, set up on different principles allowing poor opportunities for interrogation.[4]

Rapid Alert System for Food and Feed (RASFF) data demonstrates that food issues are global.[5] RASFF is used throughout the EU to communicate emerging food risks.

Between 1 Jan 2000 and 31 Dec 2008 there were (food only):

- 5034 notifications sent about food available on the market which had been tested and thought to present a serious health risk

- 1266 border rejections of food which had been tested and thought to present a serious health risk; and
- 12248 notifications of food no longer available which had been tested and thought to present a serious health risk.

The number of notifications *etc.* has increased each year since 2000. The RASFF data shows the UK to be third in the EU behind Italy and Germany in detecting issues relating to food.

Some 5% of foodstuffs are recalled owing to contamination at source, the majority of alerts happening after export checks at border crossings or during market testing.[6]

A recent check with Public Analysts shows that in the UK approximately 50,000 samples were taken in 2008, which equates to around one sample per thousand people. 25% of these food samples failed to meet legislative standards. That doesn't mean one in four meals fails to meet the standards, but is because the enforcement profession targets resources at products likely to fail to meet legislation. In 2008 approximately 12,000 food samples taken by Trading Standards Officers failed to meet the legal standard in the UK (9000 failed to meet labelling standards and 3000 failed on composition issues). Perhaps there needs to be more than general concern about the reductions in spending. In 1880 the Local Government Board (LGB) considered that at least one sample should be taken for a population of 1000. The report of the Food Products Adulteration Committee, 1896, stated that this was insufficient and that *'a local authority would do well to increase the number of samples taken from time to time until the number of adulterated samples found in those taken falls below the proportion which may be regarded as not unsatisfactory'*. A comparison of local authorities at the time led to a proposal that the sample rate should be 2 per 1000 population,[7] twice the level taken now.

Research across the EU on behalf of the Food Law Enforcement Practitioners Group (FLEP) suggests that other countries in general take more samples than the UK. In Germany, for example, they have a legislative standard which states that 5 samples per 1000 people must be achieved on an annual basis. The Germans have a similar detection rate of around 30% of samples which do not meet the legislative standards. One might conclude from this that more checks would identify more problems.

Clearly there is still a problem relating to food contamination, adulteration and labelling. The UK food sector alone is worth around £70 billion per year, so a small percentage of fraud can be worth a great deal of money.[8]

It is not true that all authorities in the UK are moving away from their commitment to fight. Many of those who run their own Public Analyst facilities demonstrate a continued commitment, one that clearly is necessary. However, working in this manner places an unfair burden on a few local authorities and the changes brought about in the legislation around 150 years ago to try and make this a more even fight are still not working. Latterly, EU legislation has introduced Food Standards Agencies (FSA) and made them the 'competent authority' responsible for ensuring sufficient levels of enforcement activity. This, together with reports by Hampton and Rodgers, has offered some local authorities the opportunity to consider if they should spend scarce resources elsewhere and leave enforcement to the Food Standards Agency. A more consistent approach across the UK is needed and perhaps even across the EU, more in accord with the Beijing declaration. This might lead to an improved ability to deal with emerging food-fraud and emergencies through improved investment and availability of new technology and the ability to match the expectations of consumers, legitimate businesses and consumers. It is time to coordinate our efforts nationally in the UK and the FSA have a key role in ensuring that this is effective.

The enforcement profession, or at least those committed to the fight, have been astoundingly good at detecting fraud and keeping contamination under control in the UK. But evidence shows that we cannot simply rely on enforcement elsewhere in the world to come up to the mark in terms of the Beijing protocol. If others cannot or will not detect their own problems then we will need to continue in the UK, even if that diverts scarce resources. Resourcefulness and a willingness to be challenged in court are two of the key attributes of Public Analysts. Resourcefulness was clearly employed by our forefathers and is still regularly employed today, even at a time of closer regulation of the science. Enforcement must not be complacent, the need for competent scientists who now have to have a wider knowledge than ever before exists. The evidence needed for the public to make informed choices about diet, evidence as described by Dame Deidre Hatton, the former Chair of the Food Standards Agency, as needing to be based on 'sound science', and the evidence needed in court, requires competent scientists who have character, tenacity and the occasional bloody-mindedness of our forefathers. It seems the fight must continue at a time when perhaps some thought the war had been won and were beginning to allocate resources elsewhere. Resourcefulness and competent scientists will be needed now as much as they ever were; the fight must continue if we do not want to be an international dumping ground for adulterated food.

REFERENCES

1. EU Regulation 882/2004.
2. A.H. Hassall, *Food: Its Adulterations, and the Methods for Their Detection*, Longmans, Green, and Co. London, 1876, p 867.
3. P.W. Hammond and H. Egan, *Weighed in the Balance, A History of the Laboratory of the Government Chemist*, HMSO, 1992.
4. C.A. Mitchell, *The Expert Witness*, 1923, W. Heffer & Sons Ltd, Cambridge, p 5.
5. T. Nepusz, A. Petroczi and D.P. Naughton, *Networked Analytical Tool for Monitoring Global Food Safety Highlights China*, School of Life Sciences, Kingston University London, open access paper.
6. The EU rapid alert system for food and feed. Available: http// ec.europa.eu/food/food/rapidalert/index_en.htm
7. J.F. Liverseedge, *The Adulteration and Analysis of Foods and Drugs*, 1932, p 6.
8. M. Woolfe, *New Scientist*, issue 2577, 15 November 2006, 40–43.